General Contracting

General Contracting

Winning Techniques for Starting and Operating a Successful Business

Mert Millman
President, CEO, Millman Construction Corporation

McGraw-Hill Publishing Company
New York St. Louis San Francisco Auckland Bogotá
Caracas Hamburg Lisbon London Madrid Mexico Milan
Montreal New Delhi Oklahoma City
Paris San Juan São Paulo Singapore
Sydney Tokyo Toronto

Copyright © 1990 by McGraw-Hill, Inc. All rights reserved. Printed in the United States of America. Except as permitted under the United States Copyright Act of 1976, no part of this publication may be reproduced or distributed in any form or by any means, or stored in a database or retrieval system, without the prior written permission of the publisher.

1 2 3 4 5 6 7 8 9 0 DOC/DOC 9 5 4 3 2 1 0

ISBN 0-07-042382-2

The editors for this book were Joel Stein and Winifred M. Davis, the designer was Naomi Auerbach, and the production supervisor was Dianne L. Walber. This book was set in Century Schoolbook. It was composed by the McGraw-Hill Publishing Company Professional and Reference Division composition unit.

Printed and bound by R. R. Donnelley & Sons Company.

Information contained in this work has been obtained by McGraw-Hill, Inc., from sources believed to be reliable. However, neither McGraw-Hill nor its authors guarantees the accuracy or completeness of any information published herein and neither McGraw-Hill nor its authors shall be responsible for any errors, omission, or damages arising out of use of this information. This work is published with the understanding that McGraw-Hill and its authors are supplying information but are not attempting to render engineering or other professional services. If such services are required, the assistance of an appropriate professional should be sought.

For more information about other McGraw-Hill materials, call 1-800-2-MCGRAW in the United States. In other countries, call your nearest McGraw-Hill office.

*This book is dedicated to my wife, Joy,
whose mystical presence gave me encouragement.*

And to "Mikey"—this author's best friend.

Contents

Preface ix

1. What Is a General Contractor? 1
2. Types of Projects to Pursue 3
3. Contracts 9
4. Legal Matters 15
5. Accounting and Bookkeeping 21
6. Insurance and Bonds 27
7. Exposure and Public Relations 31
8. Subcontractors 35
9. Responsibilities of Owners and Their Agents 43
10. Estimating 51
11. Value Engineering Changes 57
12. Schedules 63
13. Field Supervision 67
14. Preconstruction Meetings 75
15. Cost Control Records, Field Production Files, and Requisitions 83
16. Shop Drawings and Change Orders 91
17. Construction Commencement and Meetings During Construction 99
18. Quality Control and Punch List Work 105
19. Final Documents, Payments, and Performance After Completion 113
20. Potential for Growth 117
Appendix A. Subcontractor Requisition Form 127
Appendix B. Partial Release of Lien (Subcontractors and Suppliers) 129
Appendix C. Final Waiver of Lien (Subcontractors and Suppliers) 131
Appendix D. Subcontract Agreement 133
Appendix E. Requisition Printout (General Contractor to Owner) 153
Appendix F. Subcontractor Status Report Printout 161
Appendix G. Job Cost Management Printout: Budgets and Actual Costs (Labor, Material, Rental Equipment, Miscellaneous) 166

Appendix H. Job Cost Management Printout: Volume and Dollar Amounts (Labor, Material, Rental Equipment, Miscellaneous) 211

Index 281

Preface

Over the years, I have heard many young people declare their desire to become general contractors. Unfortunately, they are seldom aware of the many-faceted requirements to achieve this goal. This unfamiliarity is to be expected. How can they possess such an awareness when they lack experience in a profession so complex and detail-oriented? It is my hope that, after reading this book, they will realize that there is more to being a general contractor than picking up a set of plans and building a building. The purpose of this book is to identify, define, and explain the many elements essential to becoming a general contractor and to operating a prosperous company that will survive the high rate of failure common to the profession.

General Contracting was written not as a technical handbook, but as an informative guide and source of reference to the basic constituents inherent to the construction industry. Its primary intent is to inform young people who aspire to be general contractors, of the systems, methods, and procedures that attain the most productive and effective results in each essential aspect of the business. Additionally, it is intended to encourage and assist established general contractors to replace, or improve, nonproductive and complacent work habits. Owners, executives, all levels of management, field supervisory personnel, estimators, and office support staff may find portions of the book, or the whole, refreshingly informative and beneficial. The book can also be used as a guide for developers and lending institutions in qualifying and approving general contractors to construct their projects.

The book contains examples of unfavorable circumstances and conditions general contractors often face, and it elucidates how to minimize, avoid, or correct these difficulties with skill and perspicacity. The sequence of the chapters does not signify their order of importance. All phases of the general contracting business are interdependent—weakness in one area will be reflected in all areas. Therefore, the content of each chapter is of equivalent significance to that of another.

As a first-generation general contractor, my humble beginnings necessitated that I become exposed to all phases of the business and, on occasion, actually perform or direct every operation. This history provided me with the opportunity to gain invaluable experience. It is my most sincere hope that this experience is asserted in the book with adequate proficiency to enable general contractors to gain information beneficial to their careers. Particularly, I hope that *General Contracting* will act as a teaching tool to guide young people through easier passages in their quest to succeed as general contractors.

This book is my way of giving something back to the profession.

Mert Millman

Chapter 1

What Is a General Contractor?

General contractors are licensed professionals who agree—usually, in the form of a written contract—to construct a project in accordance with certain plans, specifications, and other related documents. Their responsibilities extend to every aspect of construction, except those items specifically designated as not included within the contract documents.

General contractors are frequently referred to as *builders*. But since a developer or owner is also commonly referred to as a *builder*, there appears to be a conflict of definition. From a practical point of view, calling a contractor a builder is a difficulty only when there is a possible confusion of the role of the contractor with that of a developer or owner. Herein lies the conflict.

Of course, there are owner-developers who perform all the functions of a general contractor when building their own projects. They may also act as general contractors when constructing projects for others. In the first instance, I prefer to refer to them as *builders,* in the second instance, as *general contractors.* All of this may be mere semantics. However, for the purposes of this book, a *general contractor* is defined as one who contracts with owners to construct the owners' projects. *Builders,* then, are owner-developers, whether or not they construct their own projects.

Emotional Requirements

I believe that being a professional general contractor is one of the most challenging and exciting careers one can pursue. It requires the emotions and temperament that allow one to enjoy the highs of success, and to endure the lows of defeat. It is not a profession for the timid, or for those who prefer a career where the sailing is smooth. Throughout construction of a project, a general contractor must be capable of making quick decisions and of dealing with unexpected changes, adverse conditions, frustration, and stress.

Competitive bidding, meeting deadlines, negotiations, and many other aspects of the profession also require a general contractor to have a competitive spirit and leadership qualities. In addition, teamwork is a prerequisite to operating a successful general contracting company. The entire staff must be tuned into each other's responsibilities, be willing and able to offer suggestions and constructive criticism, and be motivated as a unit to reach the company goals.

Recognizing the strengths and weaknesses of competitors; dealing successfully with subcontractors, owners, and consultants; and regarding every aspect of the business with strength of purpose—all are essential to a general contractor, along with the ability to enjoy the highs and endure the lows of the profession. One should combine the appropriate emotional characteristics with the stability and temperament necessary to deal—knowledgeably, firmly, and self-confidently—with those pressures common to the general contracting business.

Philosophy

The philosophy of a construction company is the foundation on which it develops. The long-term goals and financial resources of the company, and the conviction of the principals and management staff are basic to establishing these philosophies. Since concepts and theories can be redirected without changing the basic philosophies of a general contracting company, flexibility also must be employed as the company grows, or as is required by different type projects.

Established company philosophies become ineffective unless they are clearly and positively asserted to the entire staff. Once a general contractor has accomplished this, he or she must confirm that the staff understands and believes in the principles and goals of the company so as to ensure its perpetuation. Clear explications and staff commitment are the ingredients necessary to create that spirit of teamwork, enhancing positive and harmonious attitudes amongst all staff members, which is indispensable to the support of the company philosophy.

Chapter 2

Types of Projects to Pursue

Choosing the most suitable types of projects to pursue is a critical factor determining the success or failure of a general contractor. Given the proper plans, specifications, and related documents, a competent contractor should be able to build any type of building. However, this does not necessarily mean that a competent contractor can run a successful and profitable business when electing to build every type of project examined in this chapter. Selecting those projects compatible with the fundamental philosophies and long-term goals of the company and best suited to the principals and staff will strengthen the company structure. Do not feel you must build every type of project to be successful. Choose wisely.

Specializing in the construction either of one particular type of building, or of projects involving different methods of construction with diversified usage, should be the foundation on which a contractor's decisions are based. The distinctive advantage in specialization is the ease of developing and maintaining a staff thoroughly familiar and experienced with the peculiarities of the same type of projects. This serves to minimize errors in the estimating department and in the field. Most procedures become automatic, and company operations are more efficient and cost effective. Additionally, specializing is a useful way to gain the confidence of architects, engineers, financial institutions, and owners, all of whom respect one's having extensive experience and expertise in constructing specific projects. As a consequence, specialty contractors often have the opportunity to bid and negotiate projects that might not otherwise have been afforded them.

On the other hand, the advantages of being a versatile contractor, that is, one who builds a variety of projects, are equally evident. The broader scope of projects to bid and negotiate provides greater growth and profit potential. Also, versatility reduces the risk of running out of work, which would necessitate the dismissal of key personnel, and enhances a contractor's opportunity to gain a more complete understanding of the industry. The purpose of this chapter is to act as a guide that will assist general contractors in making prudent decisions when choosing the most advantageous projects to pursue.

Factors to Consider

Financial viability

Whether a construction firm is in its infancy or is an established and experienced company, the availability of financing, more than any other ingredient, will dictate which projects a contractor should elect to bid or negotiate on. It is a mistake to seek out jobs that require overhead, personnel, equipment, and other expenses that will exceed available funds or financially strain the company to its limits.

When determining which projects to pursue, a general contractor should investigate the payment schedules specified in the documents, the owner's history of making payments promptly, the retainage factor, and the amount of money required prior to payment of the first requisition. Only after ascertaining this information can a contractor effectively choose those projects that are compatible with the financial ability of the company.

Bonding

Bonding capacity, to a great extent, determines the type of projects a construction company can undertake. Obviously, a contractor with a small bonding limit cannot obtain jobs that require large bonding amounts. A contractor with substantial bonding capacity may pursue larger projects that require bonding, thereby minimizing the competition. In either event, general contractors must be cognizant that their current bonding limits may be reduced by the value of incompleted work on those of their projects not bonded.

Staff availability

Experienced staff availability, at all levels, must be considered when determining which projects to pursue. Only projects for which qualified personnel are available should be bid or negotiated. Never assume the responsibilities of a construction contract without absolute confidence that the job will be properly and efficiently staffed in all departments.

Profit/Loss potential

The reward/risk ratio should be carefully researched before bidding or negotiating a project. Certain types of projects, by their very nature, involve greater risk than others. This may, for example, apply to condominiums, whereby a general contractor is exposed to condominium association lawsuits long after the project is completed. Government work also falls into the high-risk category. Bureaucracy, red tape, excessive paperwork, and administrative details often incur expenses greater than anticipated or estimated. These examples are not meant to imply that these projects cannot be profitable, nor to suggest that limiting excessive risks applies only to these projects. Each job stands on its own. Therefore, it is essential to consider the reward/risk ratio for each project independently, despite its classification.

Competition

Competing is a way of life for a general contractor. It must be dealt with, not avoided. However, at any given time, the competition regarding certain projects will be greater than at other times. Realizing that preparing a proposal is costly and time consuming, wise general contractors will use their resources to pursue projects where there appears to be minimum competition, projects which thereby offer the maximum potential of being awarded a contract. Recognize the strengths and weaknesses of your company and of your competitors. Honestly confronting this invaluable information, will enable a contractor to make lucrative decisions when seeking new work. Contractors who specialize, often have a competitive edge when bidding or negotiating projects within their specialty.

Emotional commitment

General contractors should pursue the types of projects that the company staff enjoy building. Being a general contractor is a demanding and stressful profession. We should get as much pleasure from our work as possible. This can best be achieved by building those projects we find stimulating and satisfying.

Types of Projects

The following pages identify the major type of projects, and contain comments pertaining to each category.

High rises

These structures, whether single-purpose or multi-purpose, require specialized vertical lifting equipment, construction techniques, and unique scheduling.

One significant advantage in building high rises, is that, for the most part, once a typical floor has been reached, the remaining floors are repetitive in structural, architectural, mechanical, and electrical work items. This fact means that errors may be corrected on the lower floors, and that procedures and techniques may be improved as construction proceeds. However, a distinct disadvantage common to this type of structure is that one undiscovered mistake can also multiply itself by hundreds or even thousands of times, depending on the design and use of the building.

Government-municipal projects

These projects usually require formal proposals accompanied by bid bonds and performance and payment bonds furnished by the successful contractor. Proposals must be submitted in strict compliance with the plans, specifications, general conditions, supplementary general conditions, addendum, and other documents included in the bid packages. Normal procedure is for the bids to be opened publicly at a specific time and place.

The advantages of bidding these projects are notable. A qualified low bidder whose proposal is in order will be awarded the contract without further negotiations, provided that the bids are within budget. Construction then proceeds. Also, if the contractor's work is in compliance with the documents, payments will be made in proper amounts and in a timely manner.

The disadvantages to bidding these projects include the rigidity of the bidding and contract documents, the voluminous paper and administrative work, and the burdensome and inflexible inspections common to this category.

Industrial buildings

This work includes factories, warehouses, waste disposal plants, and other similar projects. These projects often employ unique or non typical designs requiring specialized materials and equipment. Frequently, the structural, mechanical, and electrical aspects of such structures predominate over the architectural and finishing features.

Commercial projects

These projects include shopping centers, free-standing stores, automobile agencies, restaurants, service businesses, and numerous other buildings with diverse uses.

Such projects often require many changes during construction, due to the various occupants' requirements. A general contractor must have sufficient experienced staff in order to price and process these changes as they occur throughout the construction process.

Renovations

Renovations may be the riskiest and most difficult types of project a general contractor can undertake. Usually, the plans and specifications do not fully identify and detail the work involved. This lack results in many unknown costs and time factors that are discovered only as the job progresses. Renovations can encompass major work items, such as structural, mechanical and electrical systems, or can only require simple interior changes, depending on the individual project. It is amazing what may be uncovered and discovered during construction that was impossible to foresee or estimate.

There are only two conditions under which general contractors should perform renovation work on a fixed price contract. In the first instance, the contractor must be thoroughly experienced and familiar with this type of work and thus able to anticipate and price the unknowns, including those items not defined in the contract documents. Alternatively, the nature of the job should be simple in scope, allowing the design consultants to clearly identify, detail, define, and specify all work items. In the absence of either of these conditions, a contractor should only undertake renovation work on a cost-plus-fee contract, or some other agreement that is not binding to a fixed price.

Residential work

For the purpose of this book, residential work means single-family homes, townhouses, duplexes, low-rise apartment or condominium projects, and similar structures. These buildings are usually less complicated than those previously mentioned, but they often require special skills.

On the one hand, multiple sites within one project utilize specialized crews, moving from one site to the next as their respective work items are completed. Such a method provides the most expeditious and economical procedures. On the other hand, customized homes or apartments require excellent supervision and fine quality workmanship. The general contractor's staff must then, patiently and tactfully, deal with the owners on a more personalized basis.

Chapter

3

Contracts

Different Types of Contracts

The four most commonly used contracts between owners and general contractors are lump sum or fixed price, cost plus a fixed fee, cost plus a fee based on a percentage of the cost, and upset price. The essence of all construction contracts is similar, but the peculiarities of each type reflect substantial differences concerning the responsibilities and obligations of owners and contractors. Most general contractors, at various times throughout their careers, will work under each of these contracts. For this reason, it is important to understand the advantages and disadvantages of each kind of contract and to clearly perceive its distinguishing characteristics.

Regardless of which contract general contractors execute, it is incumbent upon them to administer the same diligence and their best efforts when fulfilling their obligations. For example, the contractor's attitude and quality of performance when working with a cost plus a fee contract, whereby the owner is responsible for all costs, must be the same as when the contractor assumes these obligations, as in a fixed price contract.

Lump sum or fixed price

The general contractor agrees to perform all the work specified in the contract documents, for a fixed price. Should the cost to complete the project exceed the contract amount, the contractor must pay this cost with no recourse to the owner. There are factors that may increase or decrease the contract amount, such as change orders and other circumstances provided for in the contract.

Cost plus fixed fee

The owner is obligated to pay all costs directly related to construction, plus an agreed upon fixed sum to the contractor. The contractor's fee remains constant regardless of the final construction cost. Exceptions occur when substantial changes to the work are requested by the owner, changes that necessitate additional or less time and effort to complete the project, whereupon the contractor's fee is adjusted accordingly. Provisions that provide a basis for these adjustments should be included in the contract.

Cost plus a fee based on a percentage of the job cost

The same specifics apply to this type of contract as to a cost plus fixed fee contract, except that the fee paid to the contractor is based on an agreed upon percentage of the cost, instead of a fixed amount. A general contractor must realize that in cost plus a fee contracts, the contractor's responsibilities are, in principle, the same as in fixed price contracts. The primary difference is that the contractor does not agree to perform the work at a fixed amount or guaranteed price.

Upset price

The general contractor agrees to perform the work at a price not to exceed a fixed amount, as in a fixed price contract. The difference is that, if the total construction costs equal a sum less than the upset price, these savings either revert to the owner, or are shared by both parties at a predetermined percentage of the savings.

Experience has taught me that the fixed price and cost plus fixed fee contracts usually prove to be the most beneficial to a general contractor. The fee based on a percentage of cost contract often leaves doubts in an owner's mind. Since the contractor's fee increases with the cost of construction, did the contractor exert best efforts to keep costs to a minimum? As to the upset price contract, a contractor, in order to obtain the job, will exert best efforts to bid or negotiate the most competitive price. Therefore, the odds are greater that construction costs will exceed, and not be less than, original cost estimates.

I believe cost plus fixed fee contracts provide general contractors the most emotional satisfaction. While executing the tasks of their chosen profession, they can enjoy a greater attitude of teamwork, a greater opportunity to offer constructive suggestions and to become more involved with the project, than when working within the limited dimensions of the other kinds of contracts.

Negotiating a Contract

Negotiating a construction contract is one of the most exciting and challenging functions a general contractor performs. To efficiently and successfully negotiate a contract, contractors must be flexible and diversified in their negotiations. These techniques depend on the type of project involved, as well as the varying personalities and requirements of owners and their agents. Negotiating a contract is an art, one that requires the ability to change direction at a moment's notice, always with the goal of gaining the confidence of the people across the table. At the same time, contractors must use prudent and sound business judgment in the representation of their company.

Business knowledge

As in most negotiations, each party will attempt to secure the most advantageous agreement. Frequently, general contractors direct their primary concerns to the contract amount, time allowed for completion, and schedule of payments. In so doing, the contractor often fails to recognize lucrative vantage points that have pertinent relevance to a contract.

Percentage allowed for overhead and profit related to change orders, owner's beneficial use prior to completion, waiving of retainage on specified work items, and reduction of retainage subsequent to fixed percentages of completion are just a few examples of items that are often negotiable. Pertinacious and intelligent negotiations concerning items of secondary importance, will afford the contractor increased profit opportunity.

Technical knowledge

Negotiating a construction contract requires a general contractor knowledgeably to discuss and answer questions regarding technical aspects of the project. Experienced owners can be surprisingly astute and thorough, when inquiring about the technical issues of their project. They may wish to question and discuss the materials, equipment, and methods the contractor proposes to employ. Also, an item commonly brought up during negotiations concerns the details and intricacies of the proposed progress schedule. The general contractor must be able, with proficiency and accuracy, to satisfy the owner's technical concerns.

Value engineering changes

A general contractor must be prepared to submit and discuss these types of changes during contract negotiations, by employing an effective presentation with sufficient information. Value engineering changes are changes made to the plans and specifications to reduce the construction cost of a project (see Chap. 11). They are a valuable tool for contractors to use to further their negotiating efforts.

Basic legal knowledge

A contractor should have basic knowledge of contract law. This is particularly helpful in the case of preliminary negotiations that do not require the lawyer's attendance. When a general contractor's lawyer is present, the contractor should be capable of assisting the attorney in those areas peculiar to a construction contract. I have learned that when owners and their attorneys realize that the contractor possesses basic legal knowledge pertaining to construction contracts, a more balanced and effective form of negotiations results.

Understanding the owner

A general contractor must understand and be sensitive to an owner's needs and requirements. The owner is placing great confidence in the contractor's company and in its principals. Therefore, the owner is entitled to a contractor's attention and cooperation. Whenever possible, without causing damage or potential risk to the company, the contractor should yield to the owner's requests concerning items he or she may wish included, changed, or deleted from the contract.

Execution of Contracts

Prior to a general contractor's executing a contract, an exhausting review of the document, concerning every item and detail should be conducted. Particular attention should be given to the identification of the plans, specifications, general conditions, supplementary general conditions, addendum, and all other bidding documents that were included in the proposal. No section, paragraph, or any other detail, should be taken for granted. Even in the case where standard forms, such as AIA documents, are used, every sentence should be carefully reviewed. Often, standard printed documents contradict the intent of the agreement.

After the contractor examines the contract, and has confirmed that it complies with the terms of the bidding documents, price, and all other conditions, the contract should be forwarded to the attorney for review regarding all legal aspects. Even when the contractor's attorney prepares the contract, the contractor should undertake as diligent an examination of the document as if it were prepared by the owner's attorney.

Changes made to originally prepared contracts should be reviewed to ensure that such changes do not conflict with other areas of the agreement. Frequently, one section of a contract will make reference to other sections, paragraphs, or items. Often, a construction contract includes technical data uncommon to other types of contracts. Attorneys may not always realize that these kinds of changes can affect other paragraphs or sections of the contract. The contractor must assume the responsibility of paying particular attention in this regard.

Most government, municipal, and large corporation bid packages include a sample of the contract that will require execution by the successful contractor. These documents, especially in the case of government and municipal projects, are not negotiable, and for the most part, no changes to said documents are allowed. The contractor, when bidding these projects, must carefully examine the sample contract supplied, as this will be the document that will be executed.

One final note concerning contract negotiations, preparation, and execution. Prior to construction, owners and contractors often, with good intent, assume a passive attitude regarding seemingly mundane items. They may say, "I'll take care of that, no need to include it in the contract," or "let's not waste time on this item; you know we will do what's right." But the stresses, tensions, and numerous job problems common to the construction industry, often diminish this spirit of cooperation and the friendly attitude that are displayed during negotiations and preparation of a contract.

An experienced contractor is aware of this fact, and will spend the time and effort to have all pertinent items resolved during negotiations, and included in the contract. This procedure reduces the risk of dissension and strained relationships throughout construction.

Chapter 4

Legal Matters

It is difficult, perhaps impossible, to determine which aspect of the general contracting business is the most important. However, you may be assured that weakness in any one area will appreciably damage the company structure and potential. This certainly applies to legal representation. The attorney a contractor selects must be experienced and have a thorough knowledge of contract law, with particular reference to the laws that affect contracts between general contractors, owners, and subcontractors. One should base the choice of attorneys on their negotiation talents with respect to legal aspects of contracts, and their ability to ascertain, with certainty, that the financing provided for the project will satisfy the terms of the contract.

Due to the complex and diversified components inherent in the general contracting business, often one attorney cannot efficiently serve all the legal requirements of a general contractor. Negotiations, preparing and reviewing contracts, labor law, lien laws, bankruptcy law, litigation, corporate structure, and general counsel exemplify the areas of law in which a general contractor may require legal representation. While it may be more convenient to engage one attorney for all required legal services, it is not always the most beneficial or prudent method. The extent of the contractor's general business operations and the legal involvements regarding individual projects should be the determining factors when a contractor elects to use more than one attorney or legal firm.

Several years ago, we engaged a legal firm to act as general counsel for our company, expecting them to fulfill all of our legal requirements. Since this firm did a creditable job when negotiating and structuring our contracts, we assumed that they would be equally proficient when representing us in matters requiring litigation. This was a mistake—for reasons that will become evident.

During the course of construction on a project, one of our major subcontractors did not make proper payments, required by the subcontract agreement, to several suppliers for the materials and equipment they furnished to the project. Under the laws of the state we were working in, a statute provides for a notice to owner. This simply means that any parties supplying labor, material, equipment, or services directly to a project must advise the owner, general contractor, and bonding company, when applicable, that they will be, or are, supplying these items to the project. These notices must be submitted within a certain time frame, and in proper form, to provide protection under the law regarding payments due these suppliers.

Unfortunately, the subcontractor went out of business during construction. As the subcontractor had no apparent assets, several firms that furnished materials and equipment to that subcontractor made demands upon our company for monies they claimed due them. We had previously made payments to the subcontractor for these claimed amounts. Making duplicate payments would have resulted in a substantial loss to our company.

We advised our lawyers of the situation, and they proceeded to represent us in the ensuing litigation. After researching the law applicable to the circumstances, they advised us that the plaintiffs had properly fulfilled their obligations under the law, and suggested that we attempt to effect settlements wherever possible, and to pay in full those claims that could not be settled. They further advised us that, by continuing the litigation, we would incur additional expenses and legal fees that we could not recover.

Thinking of the amount of money at risk, we engaged an attorney whose specialty was litigation, and who was experienced in these matters. After a short meeting and review of the law pertaining to such cases, this attorney advised us that, in his opinion, the plaintiffs did not properly fulfill the legal requirements in accordance with the state statute. Consequently, he proceeded to defend our company, and the cases were adjudicated in our favor. There is no doubt that our best interest was the primary concern of the first legal firm. But, in the litigious atmosphere a general contractor works in, it is necessary to have experienced and capable representation in matters requiring litigation. When litigation is inevitable, one must not depend on attorneys who may excel in other areas of law, but who lack the characteristics and expertise imperative in a litigation attorney.

An attorney proficient in both contract negotiations and litigation can be an invaluable asset to a general contractor. This rare type of lawyer will view every contract, whether it be with an owner or a subcontractor, as a potential lawsuit. While this may appear negative and pessimistic, we must acknowledge that litigation is prevalent in the construction industry. The potential risk, when using this type of attorney, is that they frequently have a tendency to be too adamant and display excessive asperity in their attempts to better represent their client. This characteristic can be detrimental during delicate negotiations, when the attorney strays from the legal areas, and becomes involved with the business aspects of a contract.

I recall a major project we spent a great deal of time estimating, working on value engineering changes, and negotiating. Up to this point, all communications and negotiations were conducted between the owner, his representatives, our staff, and myself. When all points were finally agreed upon, we met in the office of the lawyers who represented the owner, to structure the formal legal contract. Attending this meeting were the owner, two of his representatives, his attorney, one member of my staff, our attorney, and myself. All of the business aspects that were previously agreed upon between the owner and my company were reviewed. All parties concurred to incorporate these items in the contract.

So far, no problems. Then the attorneys proceeded to verbally assail each other concerning the legal and technical points of the proposed contract. After approximately four hours of this unnecessary confrontation, it became obvious that both attorneys' prime interest was to prove their superiority as lawyers and negotiators. Eventually, their arguments drifted into the business features of the contract that were previously agreed upon. This served no purpose, and only confused matters further. Finally, I suggested that we all take a break and reconvene in one hour. During this break, I instructed my attorney to stop grandstanding, and to give careful consideration to the validity of those points he was so passionately arguing. Apparently, the owner acted in a similar manner with his attorney, because when we returned to the negotiating table, all went smoothly, and all points of the contract were agreed upon within an hour. Had the caustic attitudes and unreasonable requests of the litigating attorneys continued, I am confident it would have affected the principals involved, and the contract would not have been awarded to our firm.

These experiences emphasize how important it is for a general contractor to choose the right attorney for the job. In the case of the subcontractor who did not pay his bills, we should have chosen a legal firm specializing in litigation and experienced in representing general contractors, and not depended on our general counsel. When we needed the services of an attorney who excelled in contract negotiations, we should have looked for one whose primary background and experience was in contracts. Seldom does one attorney possess the proficiency to represent a general contractor in every area of law, or to provide all of the services enumerated in this chapter, with an equal degree of excellence.

Company Structure

Whether dealing with a corporation, partnership, or individual proprietorship, a general contractor depends on an attorney to recommend the structure that will best serve the contractor's operations and long-term goals. In this regard, attorneys will direct their primary considerations to potential liabilities.

Preparation and Review of Contracts

The essence of every construction contract between owners and general contractors is fundamentally the same. But a wide range of variables that relate to the peculiarities of each project distinguish one contract from another. Attorneys who prepare these contracts must provide equal discernment to each individual contract. The attorney's primary responsibility is to ensure that the contract provides maximum protection for the contractor despite the type of project involved or the form that is required.

Reviewing contracts prepared by others can be a more demanding task than when the general contractor's attorney drafts the contract. Every word, phrase, paragraph, and section must be carefully scrutinized for positive assurance the document reflects every item agreed to by the contractor. When subsequent drafts to the original contract are required, special attention must be directed to any revisions that have been included, to ensure that they do not adversely affect the contractor. Contract review should be performed by experienced detail-orientated attorneys, familiar with the operations of a general contracting company.

Financing

Obtaining firm evidence that funds are available for the purpose of paying the contractor in accordance with the contract documents is the responsibility of the contractor's attorney. Such a practice is not customary in government or municipal projects, but it is essential in private work.

Subcontractor Agreements

A master subcontractor agreement, designed to allow for changes peculiar to individual projects and allowing for extenuating circumstances, is prepared by the general contractor's attorney. To avoid having the agreements, in whole or in part, amended or reversed by the courts, the attorney must possess a comprehensive understanding of the laws pertaining to agreements between general contractors and subcontractors.

Negotiations

Notwithstanding business items that have direct relationship to the legal aspects of a contract, the contractor's attorneys should limit their negotiations to the legal issues. This is not to imply that lawyers should not make suggestions, and offer judicious advice to their clients regarding the business aspects of a contract. However, such advice should be given in private, prior to the preparation of the final contract.

Labor Law

Frequently, a general contractor must engage the services of an attorney adept in labor law. This area of law is usually practiced only by attorneys that specialize in labor related problems. Normally, only contractors that employ union personnel require this type of legal representation.

Insurance

Busy general contractors are exposed to an overwhelming number of insurance claims, many of which are for large sums of money. Contractors should forward all copies of correspondence, legal documents and other data pertaining to these claims to their attorney, enabling the attorney to ensure that the contractor is properly defended by the insurance company.

Litigation

Substantial sums, and other matters of importance, are usually at risk when a general contractor is a party to litigation. As shown earlier in this chapter, litigation has become a common event in the general contracting business. Therefore, only the most competent legal firms, those that specialize in this area of law, should be engaged to represent a contractor in matters requiring litigation.

General Counseling

It is advantageous for a general contractor to establish and maintain a professional relationship with attorneys who have empathy for all of the problems common to the general contracting business. While such attorneys may not possess the expertise to represent the contractor in every legal matter, they can provide the contractor with general guidance and direction.

In the construction industry, the services of attorneys are required almost every step of the way. Masterful legal representation can contribute to the success of a general construction firm in countless ways. It is essential to establish a rapport with your attorneys. They have the same emotions and feelings as other professionals and business persons. In spite of their astuteness, and the rigidity demanded by their profession, they will often provide extra energy and effort in your behalf when they believe in the people who run a company.

Whenever possible, establish an agreed-upon fee for any given service. If the nature of the required services causes this practice to be impractical or unworkable, then ascertain on what basis the fees will be charged. Such a policy can avoid dissension and strife between attorney and contractor.

Chapter

5

Accounting and Bookkeeping

Accounting

Proper and well-thought-through methods of accounting are essential to the successful operation of a general contracting company. Contractors should exercise the same discretion when choosing an accountant as when choosing an attorney. The services performed by both professionals are of equivalent importance, and are often closely integrated. Corporate structure and tax laws are merely two examples of the areas in which accountants and attorneys consolidate their expertise when selecting the most favorable methods and systems for their clients.

Accounting firms range in size from a one-person operation to international firms with offices throughout the world. The size of an accounting firm does not necessarily denote their qualification to represent a general contractor. Experience in the construction industry, a comprehensive understanding of the company's operations and long-term goals, and the ability to track the company's direction, along with the ability to supervise all the areas of accounting and bookkeeping unique to a general contracting firm are the primary considerations. Once evidence of these qualifications is confirmed, one should review the accounting firm's fees. A general contractor's accounting costs comprise a substantial percentage of the company overhead, and fees charged by different accounting firms vary by appreciable amounts. Therefore, having a clear understanding of the fee structure, and being satisfied that it is competitive, are prerequisites in choosing an accounting firm.

Tax returns

The prevalent methods of reporting a general contractor's taxable income are the completed contract method, percentage of completion method, the accrual method, and the cash basis method. The contractor is dependent on the accountant to advise which method is most beneficial, based on the firm's financial structure and long-term goals. The accountant's responsibilities extend to monitoring the contractor's operation for the purpose of suggesting changes to the method of reporting, when such changes will benefit the contractor. However, it should be noted that such changes are not easily accomplished, and must have the consent of the federal internal review commissioner.

Under the *completed contract method* of reporting, taxable income pertaining to a given project is not reported until such time as the project is substantially completed, and all funds due the contractor have been paid and all expenses determined. This is the most advantageous method for a contractor to employ, as tax payments are deferred to the end of the project. But recent federal tax legislation has curtailed some of the benefits previously afforded general contractors who used this method of reporting taxable income.

The *percentage of completion method* of reporting necessitates reporting income received and expenses paid based on the percentage of completion related to the contract amount. Since this method, to some extent, depends on estimates, the final adjustments are made when the project is completed and all receivables and payables are finalized.

The *accrual method* of reporting is used when a portion of the profit based on the percentage of completion is earned or received and the related expenses are paid or incurred. Finally, the *cash basis method* of reporting is simply reporting funds received less expenses paid. It must be understood that there is a correlation between all of the methods of reporting taxable income above, with the exception of the cash basis method.

Bookkeeping

Extensive, and somewhat complex, bookkeeping methods and systems are required in the operation of a general contracting company. The accountant will set up and update all in-house bookkeeping systems and procedures, and periodically review the work to ensure that it is being proficiently performed. The accountant also recommends the selection of computer hardware and software, and instructs the bookkeeping personnel how to use it so that the most effective results are realized.

Equipment

To a great extent, a general contractor's decisions to purchase or lease major equipment is dependant on the accountant's advice. Once the contractor furnishes the accountant with figures that enumerate the equipment cost, maintenance expense, down time amounts, and leasing expense, the accountant will consider the areas of cash flow, the advantages or disadvantages of long-term depreciation applicable to purchasing versus those of treating the leasing cost as a deductible expense, and other tax consequences.

Cash flow

Preparing cash flow projections is an important function of the contractor's accountant. These projections are depended on with respect to timing anticipated income and payables. This information is required to structure payment schedules concerning loan agreements. It also acts as a guide for anticipated construction loan draw downs. Also, accurately prepared cash flow projections will assist a contractor to determine the amount of additional work that can be undertaken without overburdening the company's cash position.

Bookkeeping

In addition to performing the mundane functions required for most businesses, a general contractor's bookkeeping department is responsible for many diverse items that are peculiar to the general contracting business. A contractor depends on the bookkeeping personnel for an updated flow of support information, and for establishing and maintaining numerous records and files. I once asked our head bookkeeper what she thought was the word most descriptive of her job. She replied, "Pressure." From that moment on, when requesting information from her department, I attempted to change my hurry-up attitude to one with a more felicitous tone.

The following records and files indicate the contractor support information that the bookkeeping department is depended upon to maintain with accuracy and continuity. Obviously, these items are in addition to the usual work performed by a bookkeeping department, such as journals, ledgers, depreciation schedules, and other files and records common to all types of business.

Job cost management files

These files reflect every financial transaction pertaining to individual subcontractors and suppliers for each project under construction. Included are the names of each firm; the date and amount of each contract; the date and amount of each bill or requisition received; and the date, amount, and check number for each payment made to them.

Dates and amounts are recorded, along with a code or brief description identifying the reason for any adjustments, such as change orders, back charges, or any other event that changes their contract price.

These files should be programmed to display updated totals for each transaction and the balance due on the contract. While the files may be structured to automatically transfer the amounts paid to the accounting journals and ledgers, their primary purpose is to provide the contractor with a quick updated financial status report of each subcontractor and supplier. (See App. F.)

Job cost budgets

Explication of these budgets appears in Chap. 10 ("Estimating") and Chap. 15 ("Cost Control Records"). The bookkeeping department establishes the codes for each work item, compiles the data furnished by the field personnel, and records it in the job cost budget files.

A bookkeeper trained in the operations of a general contractor will recognize when discrepancies between budget amounts and actual expenditures exceed tolerable limits. To avoid the necessity of scanning the entire budget, the bookkeeper will flag the work items where this condition occurs, thus saving the field supervisory staff time and ensuring their awareness. This kind of detail is only more evidence of the value of a contractor's bookkeeping department.

Subcontractors-Suppliers

A *lien* is a claim against a property or project for money due or services owing. It is the bookkeeping department's responsibility to administer the subcontractor's and supplier's release of lien files. This entails noting the date each payment is made to these firms, and checking for return of their executed release of liens that relate to each payment. When it becomes apparent that subcontractors or suppliers are delinquent in furnishing the releases, or they are not properly executed on the required forms, it is incumbent on the bookkeeping department to follow up concerning this matter. Failing to receive the releases after a reasonable length of time, the bookkeeper enters a *no further payment* note on the noncomplying firm's job cost management file. This note remains until the releases are received.

Similar files and procedures are set up for the firms that submitted notices to owners identifying the subcontractors who ordered or contracted for labor, material, equipment, and services for a specific project. Prior to making payments to the subcontractors, the bookkeeping department checks the files to ensure that releases from these firms, certifying that their accounts with the subcontractors are current, have been received. In the absence of these releases, no further payments are made to the subcontractors until such releases have been received.

Payment requisitions

The subcontractor's requests for payments, invoices for material and equipment, contractor furnished labor, services, and miscellaneous expenses are reviewed by the field supervisory staff prior to forwarding to the bookkeeper in the main office responsible for preparing requisitions. Once the bookkeeper assembles, analyzes, and codes all of the support data, the subcontractor payment requests are checked to be sure that their forms are correctly prepared, adjusted contract amounts conform with their job cost management files, previous payments in the correct amounts have been indicated, all mathematical extensions are correct, and all required supplementary documents are included and properly executed. Once these tasks are accomplished, the bookkeeper examines all material and equipment supplier invoices to ensure that charges on the invoices conform to their previously submitted price lists, mathematical extensions are correct, and required supplementary documents are included. Finally, the bookkeeper reviews, with the project manager or chief estimator, all expenses for work performed directly by the general contractor to determine the dollar value pertaining to each work item.

After the dollar amounts included in the support data have been approved or adjusted, they are entered onto the requisition form next to their respective work items. The requisition form is then checked for correctness in every respect, executed release of liens and other related documents are attached, and authorized signatures are affixed. The requisition is then submitted to the owner for payment. While waiting for this payment, the bookkeeper prepares and files all payables included in the requisition, thereby avoiding a last minute rush when payment from the owner is received. Once the owner's payment is received, the bookkeeper sends checks to the subcontractors and suppliers, makes entries in the job cost management files, and files in the job files the subcontractors' requests for payments and suppliers' invoices.

If the requisitions are prepared in the field office, I recommend that they be reviewed by the bookkeeping department in the main office, prior to forwarding them to the owner. Errors in requisitions for payment can have many adverse effects. Therefore, checking and rechecking each requisition is time well spent.

Unions

General contractors who execute contracts with the various trade unions are obligated to maintain records and files that relate to these contracts. The bookkeeping department administers these records and files, and is responsible for completing the various forms and making required payments to the unions. The bookkeeper must also verify that subcontractors performing work for a general contractor on specific projects and who have contracts with the unions, are current in their employee benefit and other union-related payments. Otherwise, the unions may attempt to collect these funds from the general contractor. These procedures are similar to those concerning firms that furnish material and equipment to subcontractors.

Accountant relations

It is important the general contractor's bookkeeping personnel work closely with the company accountant with respect to all bookkeeping and accounting responsibilities. They are expected to furnish information and documents to the accountant in a timely manner, and to display a cooperative attitude at all times. They should feel free to call the accountant for assistance and guidance as they deem necessary. These two areas of a contractor's business are closely integrated. Therefore, teamwork is necessary if favorable results are to be realized.

Chapter 6

Insurance and Bonds

Insurance

There are three areas of insurance germane to a general contracting firm: insurance required by law in order for it to operate as a licensed contractor, insurance specified in the contract documents of individual projects, and supplementary insurance deemed necessary to carry for the protection of the company. Insurance, as it relates to the general contracting business, is complex, costly, and sometimes difficult to obtain.

Many large-volume contractors with offices in multiple locales, or those with several projects in the bidding, negotiating, or construction stages will employ the services of an insurance consultant to administer their insurance needs. If this is not deemed necessary or cost effective, a contractor must then select independent insurance agencies thoroughly experienced and knowledgeable in every aspect of a contractor's insurance requirements. The insurance agencies chosen must be capable of promptly processing claims, maintaining a close surveillance over the contractor's insurance requirements, and generally fulfilling the responsibilities of an insurance consultant.

In the early stages of bidding or negotiating a project, it is the contractor's responsibility to provide the insurance agent or consultant with a copy of the specifications and other related documents that refer to insurance requirements. Shortly thereafter, the agent or consultant must advise the contractor as to the cost of the premiums, so that the contractor can include such amounts in the cost estimates. Certain insurance costs cannot be determined until the insurance company knows the dollar value of the project, sometimes required in breakdown amounts. In such cases, the contractor must be promptly notified so as to furnish the insurance company the required information as soon as possible. This entire procedure is similar to the subcontractor bidding process.

Insurance companies that insure general contractors have been known to go out of business during construction of a project. Although this is not a common occurrence, prudent contractors will place their insurance only with the most financially sound, reputable, and reliable companies. A further note, of paramount importance, is that insurance premiums are such a major cost factor that a job may be lost, or profits diminished, when a contractor fails to shop for competitive insurance rates. When pricing a job, the contractor should consider insurance premiums no less important than the cost of labor and materials.

General contractors must frequently review the subcontractors' certificates of insurance. These certificates must be submitted to the general contractor prior to the subcontractors' performing any work on a project, and under no circumstances should payments be made to subcontractors until they comply with this requirement. Each certificate should be examined for proper form, types of insurance, amounts, expiration dates, and conformance to the contract. In certain cases, the general contractor may request insurance in excess of the amounts specified in the contract documents. In this event, the subcontractors' certificates of insurance must reflect these amounts in lieu of the amounts specified in the contract documents.

Performance and Payment Bonds

Bonds are a form of insurance that guarantees the performance and payment of any work included in a contract that a contractor may fail to complete or pay for. A *bid bond* guarantees execution of a contract in accordance with the terms and conditions of the bid documents and the contractor's proposal. *Performance* and *payment bonds* are obtainable through independent insurance agencies, or in some cases, directly from the surety company providing the bonds. Depending on the type of projects general contractors pursue, the entire operation and future of the company may be contingent on their ability to provide performance and payment bonds. Therefore, it is essential that the contractor establish and maintain a good working relationship with the surety companies selected to provide bonds for the company.

As with insurance premiums, the rate structure for bonds varies from one company to another. This fact should be taken into consideration in the contractor's choice of a surety company, as bond premiums also are a substantial cost factor. Normally, general contractors will use the same surety company to provide all of their bonds. However, certain companies will charge higher rates than others, according to specific types of projects. In such cases, a contractor may elect to obtain bonds from more than one surety company.

Prior to furnishing bonds, surety companies will investigate general contractors for evidence that their contract obligations are in proportion to their financial resources, and that they have a reputation for quality performance, a history of properly paying all bills and discharging obligations, experience, and competent staff available. They will further consider the frequency and type of litigation a contractor becomes involved in, and the price spread between the bids on a project. During the course of construction, the contractor will be obligated to periodically furnish the surety company with updated information concerning percentage of completion, dollar amounts paid and received, anticipated completion date, change orders that affect the contract amount, actual cost compared to estimates, and other pertinent information. Additionally, the surety company will require copies of release of liens from subcontractors and suppliers during construction, and final waivers of liens and other documents at the end of the project. It is the contractor's obligation to comply with all of the aforesaid requirements with accuracy and in a timely manner.

When general contracting firms fail to qualify for bonds due to the company's financial status, it is possible to satisfy the surety company by furnishing personal indemnification by the principals of the firm, assuming they have assets that satisfy the requirements of the surety company. This is not the most desirable method, however, as under this arrangement, the principals become personally liable beyond the assets of the company. This is an obligation similar to personally guaranteeing a real estate mortgage beyond the value of the property. But desirable or not, this is often the only means a contractor has of obtaining bonds. Obviously, this arrangement should not be employed unless the contractor has judiciously determined that the reward/risk ratio weighs heavily in the contractor's favor, and that it is the only means to realize the company goals.

Assuming that the contract price and all other conditions of a construction project are favorable, general contractors can reduce the financial risk inherent to the General Contracting business by requiring performance and payment bonds from their subcontractors. Bonding all of the subcontractors will provide maximum protection. However, bonding only the major firms is usually adequate, and more practical.

Before young people aspiring to be general contractors breathe sighs of relief and assume that bonding subcontractors is an easy solution, they should realize it is not an easy, nor always practical, arrangement. Because bonding is a costly item, this additional expense will increase the amount of the contractor's proposal. This is a particular concern when the owner only requires a bond from the general contractor, and not from the individual subcontractors. In this event, the additional bonding expense, included in the contractor's cost estimates, may reflect a higher bid than submitted by contractors who do not bond their subcontractors. Additionally, proving to the satisfaction of a surety company that subcontractors have not fulfilled their obligations under the terms and conditions of their agreements is sometimes a difficult and costly task.

Chapter

7

Exposure and Public Relations

Throughout the United States there is no shortage of general contractors. Competition prevails for every type of construction project, and the road to success is lined with contractors eager to succeed. Effective exposure and public relations afford a contractor the opportunity to compete. Simply stated, let the public know who you are, and what you do. But, as in all aspects of the general contracting business, exposure and public relations must be well planned. Tell the story of your company, its principles and philosophy. If specializing is company policy, proclaim the type of project and expound the company experience and achievements confidently. Such confidence is infectious.

I believe that general contractors should only employ the highest standards of exposure and public relations, standards appropriate to professionals who share in the responsibility for the landscape of our universe and the safety of its inhabitants. I do not subscribe to self-serving pretentious means of reaching the public through television, newspapers, or other types of media. Treat your company with the respect and dignity it has earned—the public will respond accordingly.

Project Signs

The primary purpose of project signs is to identify the general contractor on a construction project. Signs also serve to keep the company name before the public. These signs should be consistent—in their colors, copy, type style, and art work—with the company letterheads, vehicles, and equipment. Copy should be limited to the company name, cities where offices are located, and phone number. More than this is superfluous, and will diminish the value of the project signs. Many prominent and successful contractors use only the company name on their project signs, a practice which emphasizes the importance of keeping the signs uncluttered and simple.

Project signs must be clean, freshly painted, constructed of durable material, and erected in a secure manner. When the size of the signs is not specified in the project specifications, good judgment should determine the size appropriate for different projects. Project signs should be erected in the most visible locations that require the minimum amount of relocating as construction progresses. Other signs to be erected on the sight identifying subcontractors, design consultants, lending institutions, and others should be placed in locations that will not detract from the general contractor's project sign.

Brochures

Brochures are an excellent medium for making known the essence of a general contracting firm. They afford a contractor the opportunity to tell the company story through copy and illustrations. Introducing and explicating the experience of the principals and management staff, illustrating and identifying company-built projects, making known construction techniques unique to the company, and expounding on the principals, background, and goals of the company—all these are relevant items to include in a general contractor's brochure.

Brochures, to be effective, must be designed to serve the purpose for which they are intended, that is, procuring new clients. To successfully accomplish this purpose, prospective clients, architects, engineers, and others related to the industry must find the brochures informative and interesting. Copy should be concise and to the point, and not overemphasize the message; illustrations should be of good quality and not cluttered; the overall design should be strong, but not ostentatious. And, please, do not incorporate photos of company executives dressed in suits, ties, and hard hats pretending to be studying a set of plans on the job site or in their office. Let the brochures be straightforward and tell a truthful story. Remember, the people for whom they are intended are successful, experienced, and knowledgeable. They will not put much credence in brochures that are cluttered with soap opera techniques.

Distribution of the brochures should be administered with the same degree of importance afforded their design and contents. Unless they reach the market for which they are intended, they serve little or no purpose. In the design stage, the brochures should be constructed in a manner that easily allows adding in a later printing, new photographs and copy of interesting new projects.

Publications

Newspapers and industry publications are excellent media for making the public aware of new construction contracts awarded to a general contractor. Other events such as a contractor's development or employment of new techniques of construction, the opening of offices in new locales, and other interesting or unusual items pertaining to the company or its principals can be avenues of public relations when printed in local newspapers or industry-related publications. Engaging a public relations firm experienced in the operations of general contractors to develop and submit the copy to the various publications usually produces better results then when it is submitted directly from the contractor's office.

Photographs

Color photographs or renderings of interesting projects built, or being built, by the general contractor should be displayed in the areas most visible to prospective clients. Such images make an impressive statement, and add credence as to the contractor's experience and qualifications. Reference to a photograph or rendering that has design and usage characteristics similar to projects owners intend to build is an auspicious way of capturing their interest and gaining their confidence.

Used in conjunction with the company brochures, these photographs and renderings tell a more credible story than would lengthy commentary by the contractor. At this point, owners know that the contractor is capable of constructing their building. Therefore, the contractor need only elucidate those qualifications not identified in the photographs and brochures. General contractors are often more proficient in managing a construction company than in proclaiming its virtues. Letting the pictures do the talking is sometimes the most effective type of salesmanship a contractor can employ.

Letters of Reference

Complimentary letters from owners, design consultants, and lending institutions are the least self-serving and most potent public relation tools a general contractor can possess. They should be solicited, when possible, and used for the purpose of gaining the confidence of prospective clients.

Main Offices

Attention to the public relations aspect of a general contractor's business should be evident in the company's main office. Adequate room and conveniences should be provided for owners and their agents who attend meetings in the contractor's office. Areas used by the owner should be efficient in their design, but have a relaxed, pleasant, and friendly atmosphere. When observing the estimating, bookkeeping, and other departments, the owners should be satisfied that they are of adequate size, effectively laid out, properly equipped, and functional in every respect, to ensure that their projects are receiving the attention from the main office support staff they deserve.

Field Offices

It is important to keep the field offices orderly and clean, and to provide ample room for owners and their agents when they attend meetings in these offices. Every operation of a general contractor is scrutinized by owners. Field offices are no exception.

Chapter 8

Subcontractors

Approximately 85 to 90 percent of the dollar value of work on a construction project is performed by subcontractors. The balance is accomplished directly by the general contractor. These percentages vary depending on what scopes of work a general contractor chooses to perform, using company labor, material, and equipment.

Realizing that subcontractors perform the major part of the work, the general contractor *must* use keen judgment when selecting subcontractors for each project. Many general contractors have a list of preferred firms to which they award the majority of their work. They are more confident of better performance and less risk when working with familiar and proven subcontractors. Assuming that these subcontractors' prices are competitive, it is only logical for the contractor to follow this pattern.

There are two distinct disadvantages a general construction company faces when it habitually gives its work to the same firms. First, the company may not receive the lowest or most competitive bids from competing subcontractors. These firms realize that, all things being equal, the subcontract agreement will probably be awarded to a firm on the contractor's preferred list. Also, they are aware that if they do submit the lowest bid, certain contractors will offer preferred subcontractors the opportunity to meet the price, terms, and conditions of the "lowest" proposal. When more than one general contractor is submitting a proposal on a project, it is to be expected that the subcontractors will quote their lowest price to those contractors that do not show partiality to favored subcontractors. The second disadvantage is the strong possibility that subcontractors will not administrate their best efforts on projects for a general contractor who only rarely gives them a job. It is likely that their most skilled and experienced labor force will be employed on projects where the general contractor affords them equal consideration. In order to receive the most comprehensive and competitive bids, a general construction company must make conscientious efforts to establish working relationships with several subcontractors related to each scope of the work. Making a practice of using different firms for various projects will prove advantageous over the long term.

An experienced general contractor will investigate several factors crucial to a choice of subcontractors before entering into subcontract agreements. The factors listed in the following paragraphs are not in order of importance; they are of equal importance. (See App. D.)

Precontract Problems and Solutions

Bid price

Obviously, in order to merit consideration, a subcontractor's price must be competitive. However, when multiple subcontractors submit proposals that reflect large price discrepancies, the low bids must be carefully reviewed to ensure that all required scopes of work have been included, and that the low bidder qualifies in every respect. This review must be accomplished prior to including the low bid in the general contractor's job cost estimates.

Scopes of work

This item can be deceptive. Work items a general contractor may assume will be performed by certain subcontractors are not always clearly identified in the plans or sections of the specifications that apply to these subcontractors. They may be referred to in the general conditions, supplementary general conditions, or addendum, but without relation to specific subcontractors. Subcontractors requested to bid on a project should be experienced enough to recognize industry-wide performance expectations even when work items are not specified in the bidding documents, and conscientious enough to include those items in their bids.

Staff

Qualified subcontractors will have sufficient and experienced staff in all departments—management, estimating, and, most important, supervisory personnel. Such a requirement applies to unusual projects needing field supervision and personnel familiar with the peculiarities of these projects. Progress and quality workmanship are impeded when subcontractors are unable, or refuse, to furnish suitable personnel to a project. Regardless of a subcontractor's other favorable qualifications, the general contractor, prior to execution of a subcontract agreement, must be assured that the subcontractor has the ability and the will to comply with these requirements.

Financial viability

Subcontractors are only dependable to the extent of their financial viability. Lack of funds to make payroll, to purchase materials and equipment, and otherwise to support the firm between payment requisitions will prove disastrous to a subcontractor's business. In the case of prime subcontractors, these disastrous results may extend to the general contractor. The temptation to award a contract to a subcontractor who is a low bidder and qualifies in all respects except financial ability must never be yielded to. The financial soundness of a subcontractor must be carefully considered.

The bonding capacity of subcontractors is governed, to a large extent, by their financial strength. There are many subcontractors who do not have the financial structure to qualify for bonds, but are strong enough to fulfill their obligations in every respect. Many construction contracts do not require subcontractor bonds, and in these instances, a general contractor may elect to use a subcontractor who cannot provide a bond, but qualifies in all other areas, including lowest bid, past performance, and general credibility. Although this may be an advantageous and prudent decision, the general contractor must first examine the potential risks that can evolve from not using a bonded subcontractor. The general contractor must review the subcontractor's current financial statement and other work under contract. If the results are favorable, and the contractor chooses to use a nonbonded subcontractor, financial protection can be reinforced by restricting, when acceptable, new work the subcontractor may undertake until a certain percentage of work in progress is completed.

Workload

The nature of the construction business is such that it is impossible for subcontractors to determine, with accuracy, when they will have new jobs under contract. This intrinsic characteristic of the industry frequently compels subcontractors to take new work as it becomes available. Often, the results are workloads beyond their capacity to perform properly on new projects. Prior to awarding a subcontract agreement, the general contractor should be assured that the subcontractor's current workloads will not impede the ability to perform efficiently when undertaking additional work.

Value engineering changes

A competent subcontractor will have sufficient knowledge to suggest value engineering changes that are neither frivolous nor inconsistent with the basic concepts and intent of the project plans and specifications. Effective value engineering changes proposed by a subcontractor can assist general contractors to prepare a list of changes for submittal to an owner. This skill should be considered when selecting subcontractors.

Attitude

Positive and cooperative attitudes of the entire subcontractor's staff are imperative to the timely and efficient production of a construction project. Despite all the proficiency and qualities a subcontractor may possess, negative and nonproductive attitudes on the part of the principals, executives, and job supervisory staff is cause for the firm's disqualification. A general contractor's staff will be burdened with excessive waste of time and stress when working with subcontractors who habitually and unjustifiably display contrary and belligerent attitudes. Each scope of work on a construction project interlaces with, in some manner, and depends on other scopes of work. Therefore, successful performance requires the cooperation and helpful attitude of all subcontractors.

To help ensure the selection of subcontractors displaying productive attitudes, a general contracting management staff should meet with the subcontractor's principals and executives prior to execution of a subcontract agreement. Since attitudes cannot effectively be included in a legal document, these personal investigations are necessary. Also, general contractors should heed the advice of their project superintendents who have had prior experience with the subcontractor's field personnel.

It is not the intent of this chapter to enumerate each possible problem of individual subcontractors on a construction project regarding performance and quality of every work item. However, a general contractor must be aware of the five most generalized subcontractor problems that occur during construction, preventive courses of action, and solutions.

On-the-Job Problems and Solutions

Job progress schedules

The most common problem subcontractors have in complying with schedules is simply their lack of understanding of, or attention to, the schedules. The first step in preventing this problem is for the general contractor to include in the agreement the subcontractor's obligations to perform in accordance with the job schedules prepared by the general contractor. This article must be stated in clear and concise language, and treated as a prominent part of the agreement.

To ensure that the subcontractors understand and believe in the schedules, the general contractor's supervisory staff must monitor the work daily, or as often as required. Immediately upon recognizing that certain work items are falling behind, supervisors must review the schedule with the responsible subcontractor's superintendent, for the purpose of ascertaining the causes of the delays. If delays result from the subcontractor's lack of understanding of, or attention to, the schedules, the contractor must explain and interpret those items not clearly understood, and strongly request the subcontractor's closer attention to the schedules. Repeating this procedure as often as necessary will appreciably diminish unwarranted delays in job progress.

If this method proves to be of no avail, the best solution is for the general contractor to request replacement of the subcontractor's supervisory personnel with people that both understand and are willing to give proper attention to the job progress schedule.

Substandard quality

Subcontractor failure to perform work with the standard of quality demanded by the contract documents and the standards of the industry is one of the most common and devastating problems a general contractor must deal with. It is time-consuming, costly, and damaging to the general contractor's reputation.

The easiest way to mitigate this problem is the simplest. Only use experienced subcontractors, who recognize and respect the quality standards required by the contract documents. The prevention, or more probably, the minimizing of the problem is accomplished by constant field supervision of quality control. Extremely close surveillance should be furnished as each subcontractor commences work. Before allowing the subcontractor to proceed, the work should be inspected for quality defects. As soon as defects appear, proper corrective procedures and methods should be immediately implemented and employed throughout construction. Follow-up quality control inspections in every category of work should take place throughout construction. Preventive and corrective measures must be immediately activated as deemed necessary by the inspecting personnel. Subcontractors are more likely to produce acceptable quality when aware of intense supervision of their work by the general contractor.

When unacceptable quality does occur, the most efficient solution is to have the defective work corrected without delay. Since each work item in construction depends on a preceding work item, it is the responsibility of each subcontractor to perform the necessary corrections prior to the start of subsequent work. This will serve to avoid time delays in job production and substantially reduce cost of repairs. An additional solution to having substandard quality work properly and expeditiously corrected, is to withhold from the subcontractor's requisitions for payment the money required for such corrections, until satisfactory corrections have been accomplished. When employing this method, the general contractor must use prudent judgment, as withholding too much or too little from a subcontractor's payments can be counterproductive.

Shop drawings and submittal data

Late subcontractor delivery of shop drawings, samples, and other required data causes delays to a project. Incomplete or improperly prepared shop drawings can result in costly mistakes in the field.

These problems can be avoided by early notice, to subcontractors from the general contractor, defining the specific time their various data are required to be submitted. This time frame must allow for the general contractor's review and corrections, and errors and deficiencies in the submittal data must be promptly brought to the subcontractor's attention. Follow-up by the general contractor of all aspects of subcontractor-furnished drawings and data must be accomplished frequently.

When all preventive measures have failed, the only practical solutions are to withhold payments until approved shop drawings and related data are received, and to reduce the subcontractors' contract amount by the damages resulting from their failure to perform.

Requisitions for payment

It is not uncommon for subcontractors to request payment in amounts exceeding the value of work they have performed. This problem, when not corrected, can arouse suspicions of the general contractor by the owner, and even more serious, result in the subcontractors' not having sufficient funds to complete the work included in their contract. Additionally, when subcontractors are overpaid, or paid in advance, it may become more laborious to have their work performed and completed in a timely and proper manner.

Prior to receiving subcontractor's first requisitions for payment, a schedule of payments in breakdown form covering the value of all work included in the subcontractor's agreement should be submitted to the general contractor for review, corrections, and approval. Upon receipt of the subcontractor's requisitions, the general contractor must measure the value of work performed as reflected in the approved schedule of payments. These procedures will help prevent overdraws by the subcontractors.

When subcontractors habitually request excessive payment, not in accordance with the approved payment schedules, the general contractor should call a meeting with the subcontractor's top management personnel, for the purpose of establishing firm rules and standards concerning future requests for payment. Should the problem continue, the most effective solutions are to reduce the amounts of the requisitions accordingly, and to serve notice of noncompliance.

Punch list work

The most serious problem in this scope of a subcontractor's corrective work is the failure to execute in a timely and orderly manner. The most intense inspections and scrutiny by the owner's representatives is given the project while the final punch list work, is being performed. Failure to complete properly, and on time, the punch list work will delay final payment, and damage or destroy an otherwise amicable relationship between the owner and general contractor.

This problem, in some instances, may also constitute failure to complete a project within the time allotted in the contract. When penalty clauses related to completion dates are part of the contract, financial losses to the general contractor can result. Additional supervision, overhead, and other costs will escalate when subcontractors do not complete their punch list work within the prescribed time.

To help prevent these occurrences, a general contractor must commence inspections related to subcontractor punch list work at the earliest possible time. Corrective work as required by the subcontractors should begin immediately upon substantial completion of an area or work item, and be continuously performed until the work is satisfactorily completed.

When subcontractors fail to complete their punch list work in a timely and acceptable manner, a general contractor, after serving proper notice upon the subcontractor, should perform the work employing company labor, materials, and equipment, whenever possible. When this is not possible or practical, the contractor must employ others to complete this work. In either event, the cost should be deducted from monies due the responsible subcontractors.

Chapter

9

Responsibilities of Owners and Their Agents

In this book, owner's agents are recognized as any person or firm representing the owner concerning construction of a project. Throughout the book, references to the owner include the owner's agents. Agents comprise personnel directly employed by the owner and such independent firms as architects, engineers, and other design consultants.

The preceding and subsequent chapters identify and explain the diverse items and responsibilities a professional general contractor is depended on in order to perform excellently and propitiously. Equally important to a successful construction project is the determined and attentive performance of the owner. Always be aware that the owners are a party to the contract, and, as such, must fulfill their obligations with equal diligence and effort as the general contractor. If the project specifications or other documents do not clearly define owner responsibilities, the general contractor must ensure they are included, and clearly identified, in the contract.

I recall three specific experiences, concerning different projects, that exemplify owners' attempts to avoid or negate their obligations. The first relates to the foundation work of a large project for which the contract specifications unambiguously stated that the owner was responsible for preparing the subsoil under the building. This preparation included compaction to elevations specified on the drawings. In the early stages of construction, we discovered that, in certain areas, the subsoil work previously performed did not conform to the specifications. In a meeting about this matter, we elucidated the problems to the owner and referred to the appropriate portions of the contract documents governing this matter.

The owner, an experienced builder, clearly understood the problem.

Despite acknowledging that soil preparation was not our responsibility, he demanded that we make the necessary corrections without any additional cost to him. He assumed, taking an oversimplified and erroneous position, that an owner should not be burdened with these extra charges. To further complicate matters, the contractor who had performed this work for the owner was no longer in business. After much aggravation and loss of time, the matter was finally resolved, whereby we performed the additional work and were paid for the extra direct job costs, but were not compensated for delays, additional overhead, or lost profit.

The second project had more diversified problems due to the exorbitant amount of errors, omissions, and discrepancies in the architect's working drawings. Such problems would have been extremely difficult, perhaps impossible, to discover during the course of estimating, particularly within the time allowed to prepare a competitive formal proposal. Review of our estimates revealed that members of our estimating staff performed their work in a detailed manner, and within the dictates of the bidding documents.

The errors, omissions, and discrepancies began to affect the quality of work, create confusion, and delay job progress. As should be expected of a professional general contractor, we made every effort to solve the problems in the field, but the architect refused to cooperate with our field staff; it became evident that he intended to avoid his responsibilities. Finally, we requested a meeting in our office with the architect to obtain clarifications and to circumvent similar problems throughout construction. At this meeting, we clearly defined the problems we were facing, and requested his assistance. We pointed out that, at considerable cost to our company, we had made several changes in the field—changes necessary to correct his errors—without requesting additional payment from the owner. But we advised him that this practice could not continue, as the cost and time delays were becoming excessive. Additionally, we were concerned that, in spite of our good intentions, we could be in violation of the contract by making changes without formal recordation and approvals.

The architect acknowledged the problems, but advised us that he drew the working plans to the best of his ability and, therefore, would assume no responsibility for his admitted errors, omissions, and discrepancies. He insisted that we neither advise the owner of the problems nor request extra payment for the necessary corrections. The meeting failed in its purpose. We promptly documented all of the required changes, the reasons for such changes, and their cost. This information was submitted to the owner with our request for a formal change order. As a result of subsequent time-consuming meetings, we did receive payment for the additional work. However, the time, cost, and strained relationships could have been avoided had the architect recognized and fulfilled his obligations.

The third experience further emphasizes the importance of design consultants' willingness to acknowledge and acquiesce to the disciplines of their responsibilities. The structural system of this project was cast-in-place reinforced concrete. The design employed unusual configurations in certain areas of the floor slabs, requiring utmost attention to the concrete form work, installation of the reinforcing steel, and placement of the concrete. As experienced general contractors in this type of work, we provided the project with our most skilled supervisory personnel, who were familiar with these unique design characteristics.

Much to our dismay, upon removal of the forms in one area, our inspection revealed excessive stress cracks and a potential structural failure. Further work related to this scope of the project was immediately stopped while inspections and investigative measures were initiated to determine the cause of the problem. Upon completing exhaustive research and investigation into the methods and materials used in the construction of the problem area, we concluded that all of our work, and that of our involved subcontractors, was performed in a proper manner and in strict accordance with the plans, specifications, approved shop drawings, and all other pertinent data. We conveyed this information, including progress photographs, job logs, inspection reports, and results of samples and tests, to the engineers who designed the structural system.

The engineers claimed that they had checked their design and found it to be adequate in every detail. Therefore, they surmised that the fault was in the methods or materials used in construction. They insisted that we stripped the concrete forms too early, improperly installed the reinforcing steel, or used concrete of a quality or strength not consistent with their specifications. Although a field inspector employed by these engineers inspected and approved the erection and removal of the concrete forms, installation of the reinforcing steel prior to concrete placement, installation of sleeves in the forms for future electrical and mechanical installations, and placement of all electrical, plumbing, and mechanical piping, the engineers insisted the fault was ours and so advised the owner. Further tests and investigations— including additional concrete cores and x rays—were performed, only to reinforce our position that the materials and methods used were in compliance with the plans and specifications.

As a result of the conclusive evidence compiled from the foregoing investigations, we suspected the peculiarity of the design in the problem areas was not compensated for in the engineer's drawings. This conclusion became significant when inspections disclosed no problems in the areas contiguous to the problem area. Once again, we requested the engineers to recheck their calculations. They denied our request, and continued to disclaim any responsibility concerning this matter. At this point, we requested a meeting with the engineers and owner.

We suggested that the owner engage an independent engineer acceptable to him, the designing engineers, and our company for the purpose of investigating and determining the specific reasons for the defects. The owner agreed. An engineer of national acclaim in the field of concrete construction was engaged for this purpose, as well as to recommend corrective procedures. After reviewing the design drawings, he concluded that the reinforcing steel was improperly designed for the nontypical areas where the defects occurred, and recommended necessary revisions. Most of the lost time and cost would have been avoided, had the engineers spent more time checking their work and less time denying responsibility for the problem.

The intent of relating the foregoing experiences is not to imply that the majority of owners and design consultants do not satisfactorily fulfill their obligations and responsibilities. On the contrary, most knowledgeable, experienced owners and consultants realize the importance and magnitude of their responsibilities. They perform in an efficient, professional, and commendable manner. But similar problems can and do occur during the course of a construction project. General contractors, realizing that their gross margin of profit is small and their risk is great, must be able to deal with adverse circumstances and to assert themselves even in the face of owner-caused problems. When contractors sagaciously gain an owner's confidence, show evidence of cooperation, and relate their position with professional acumen, such problems can often be resolved without dissension or loss to innocent parties.

Several of the important owner functions and responsibilities related to a general contractor's work are explained in the succeeding paragraphs of the chapter. Contractor assistance to the owner and specific methods of avoiding problems are included.

Construction Documents

Before beginning construction, it is the owner's responsibility to have all revisions to the contract incorporated in the working plans, specifications, addendum, and related documents, and to provide the contractor with revised drawings, details, and documented information required throughout construction. When practical, the addendum items, bulletins, and instructions issued prior to contract execution, should be included in the plans and specifications. This practice eliminates dependence on notes and memory, and helps avoid mistakes in the early stages of construction. Preventing paradoxical clarifications and details is best accomplished when the contractor's staff carefully reviews all revisions for correctness and for the way they are incorporated in the contract documents.

Frequently, completed interior design drawings and details for public areas are not furnished to the contractor until substantial work has been performed. For the purposes of coordination, continuity, and avoiding delays in job progress, the owner must submit these plans and details to the contractor at a time consistent with job progress. It is the contractor's obligation to provide the owner with the specific dates that such plans and details will be required, allowing time for review, pricing, processing, and distribution. Follow-up by the contractor is imperative, as owners often procrastinate in their selection and approval of interior design items.

Shop Drawings and Samples

The review and processing of shop drawings and samples submitted by the general contractor is an important owner responsibility. Whether the owner assigns this task to the design consultants or other representatives, the shop drawings and related data must be properly checked and returned to the contractor without undue delay. Assistance to the owner, in this regard, requires three fundamental contractor actions: furnish a schedule indicating the latest acceptable dates for return of submittal data to the contractor, allow reasonable time for the owner's review and processing, and check and correct data prior to forwarding to owner.

Color and Material Selections

When these selections are not specified in the contract documents, it is the owner's responsibility to furnish this information to the general contractor by the requested dates. As in the case of interior design drawings, owners and consultants are sometimes delinquent in making their final selections concerning finish items. Providing the owner with the required dates in documented form and, when practical, applying color and material samples to designated areas of the building, often serve to assist owners in their selections and in advising the contractor within a time frame consistent with job progress.

Change Order Work

The delay of decisions and direction regarding changes to a project under construction is one of the most prevalent causes of job progress delays. When a general contractor submits prices and information in proper form and detail, it is the owner's responsibility to respond promptly. This responsibility extends to delays and to cost in excess of the contractor's original estimates, when owners fail to process change order requests within a reasonable time frame. It is often necessary for a contractor to follow up concerning owners' decisions regarding pending change order work.

Inspections and Testing

Owner-furnished inspection and testing obligates the owner to employ firms experienced and proficient in this work. The owner is responsible for timely performance, accurate reports, and proper documentation. Once the inspectors and testing laboratories have been selected, the contractor should review the systems and methods to be used, thereby avoiding future misunderstandings and problems.

Meetings

Meetings among general contractors, owners, and consultants are necessary throughout a construction project. These meetings serve as the best means to maintain communication, provide clarification and direction, and find solutions to job problems. They are especially beneficial when misunderstandings and conflicts cannot be resolved through other means of communication. It is the owner's responsibility to attend these meetings as reasonably requested by the contractor.

General contractors must present their request for clarification, information, and direction in a definitive and orderly manner. Such a presentation contributes to receiving the owner's cooperation and helps to avoid a combative atmosphere. Since owners and their agents are busy, meetings should be scheduled well in advance, and at a time convenient to all parties. And owners, aware of the value of the contractor's time, should expedite furnishing the information and assistance requested by the contractor.

Owner Use Prior to Completion

Owners must advise the general contractor of the specific areas they intend to use for offices, models, marketing, or other purposes during construction. Adequate preparation time for these areas so as to avoid work interruption usually requires the owner to furnish this information in the early stages of construction. When using designated areas of the building prior to completion, it is the owner's responsibility to adhere to the contractor and governmental safety rules and regulations, and to reimburse the contractor for any additional cost related to this work not included in the contract.

The owner and contractor should meet before construction starts, or as soon thereafter as possible, to review the conditions and extenuating circumstances applicable to this item. Times of completion, cost projections, and other pertinent issues should be established and agreed upon. This should provide solutions to what otherwise would be latent problems.

Requisitions for Payment

Notwithstanding the importance of the foregoing owner responsibilities, perhaps the most important is the prompt and proper payment of the general contractor's payment requisitions. Whether inspections and reviews of the requisitions are performed by one person or by multiple parties, it is the owner's obligation to expedite the process enabling timely payment. The owner must be made to realize that construction progress is contingent upon the contractor receiving payments in accordance with the terms of the contract. On the other side of the ledger, the contractor should only request payment for work actually accomplished, submit requisitions by the prescribed dates using approved forms, include all specified affidavits and releases, and fulfill all contractual obligations conscientiously.

The significance of cooperation and teamwork between owners, their agents, and the general contractor cannot be sufficiently emphasized. These are truly the inescapable responsibilities of owners and general contractors, and they are difficult to define clearly in a contract. When both parties acknowledge and respect these responsibilities, many problems common to a construction project will be circumvented.

Chapter

10

Estimating

Estimating is one of the dominant forces from which a general contractor's business is developed. Fundamentally, the initial responsibilities regarding new projects are delegated to the estimating staff; their work product determines whether or not a contractor will be the successful negotiator or bidder. The potential profit that will be realized from a construction contract is dependent on the accuracy and astuteness of the estimating personnel. Hence, the future of a general construction company is contingent upon the professional acumen of their estimating department.

Estimating is very tedious work, and requires the utmost attention to details and technicalities. Addendums, bulletins, notices, checking and rechecking, revisions, time schedules, meetings—all are evidences of an estimator's work. Efficacious estimators have three basic qualities: experience, proficiency, and tenacity. Understanding and visualizing how a project will be constructed in the field is an expedient quality of good estimators. Therefore, excellence in the preparation of a job estimate will more probably be realized when estimators have actual field experience. Observing methods and procedures on a job site is an invaluable way to inspire the growth and maturity of a construction company's estimating staff.

When practical, it is advantageous to assign two or more estimators to a project. This enables them to better utilize their experience and expertise in those areas in which they excel. When employing this method, it is of utmost importance that the entire staff work as a unit, fully aware of all segments of the plans and specifications. Since all work items of a construction project are dependent on each other, each estimator must consider all the ramifications that directly or indirectly affect the items they are estimating. At least one staff member, most commonly known as the chief estimator, must possess the discernment to coordinate, schedule, evaluate, and supervise the entire preparation of a job estimate.

The scopes of work required to prepare a comprehensive estimate for a construction project are identified in the subsequent paragraphs of this chapter. Common problems and their prevention are included.

Review of Plans and Documents

Preparing a job estimate begins with a concentrated review of the plans, specifications, general conditions, supplementary general conditions, addendum, and all other related data. Notes and memos of any unusual conditions or circumstances are recorded. For the purpose of avoiding errors and omissions, these reviews should be repeated at frequent intervals throughout the estimating process. This is significant when the plans and specifications are extensive and complex.

Recap and Spreadsheets

These worksheets itemize all scopes of work to be performed by the general contractor and subcontractors. The project specifications may serve as a guideline, however, they cannot be depended on to be all-inclusive. The spreadsheets are structured in breakdown form, indicating the costs for labor, material, and equipment for each work item. The sum total of these breakdown amounts are entered on the recap sheets next to their respective scope of work. Adjustments in price and breakdowns will occur throughout the estimating process, therefore, the recap entries should not be recorded until final reviews of the spreadsheets have been completed.

Quantity Estimates

This is the most time-consuming and exacting work performed by the estimating department. Only experienced personnel who are perceptive and patient should be assigned to this task. Quantity estimates are completed for all work performed directly by the general contractor that requires labor, material, equipment, rental equipment, and miscellaneous items. These estimates must be accomplished in a systematic manner, with sufficient breakdowns for each item and area to enable reviewing and preparation of cost budgets. Reference to page numbers, sections, and details of the plans should be noted next to each takeoff item on the forms used to record the quantities to facilitate checking and cross referencing. It is advisable to record the cost for each item, after all the quantities have been determined, thus affording the estimator a more complete understanding of the project prior to pricing.

When time allows, and staff is available, quantity estimates should be made of the work to be performed by the prime subcontractors. This affords the general contractor greater opportunity to efficiently compare competitive subcontractor bids and to negotiate more favorable subcontracts. These breakdowns are also instrumental when checking subcontractor prices related to change order work during construction.

Computers and high-tech equipment are valuable tools for making quantity estimates. They eliminate errors, save time, and produce a better overall work product. Computer and software selection should be performed by consultants knowledgeable in the workings of a general contractor's estimating department. In order to realize the most productive and auspicious results from computers, the estimating staff must be thoroughly trained in their use. Computers for a professional competitive contractor are a necessity; they are no longer an option. Their importance merits intelligent selection and intensive use.

Bid Assembly

Upon determining the subcontractors and suppliers that will submit bids, the estimating department must promptly advise the firms the due date of their bid, and when and where they may secure the plans and specifications for estimating purposes. Individual forms should then be created, using the same codes or work item numbers indicated on the recap sheet and including each bidding firm's name, address, date of contact, and scope of work. Space for breakdown items and their respective prices; the dates that bids are received; remarks; and total price, including applicable taxes, should be provided for on the forms. This method of recording bids allows the contractor to review all competitive proposals in an orderly and efficient manner. Early preparation of the forms avoids confusion and allows time for whatever adjustments may be required throughout the estimating process.

Meetings in the general contractor's office should be scheduled with the bidding subcontractors and suppliers for the purpose of reviewing the bidding documents, scopes of work, clarifications, items to be included or deleted, and all other related items pertinent to the project. All unusual circumstances that may affect the subcontractor's or supplier's bids, as determined at the meetings, should be recorded on the individual bid forms. The general contractor's estimating staff is obligated to promptly furnish copies of all addendum and other information received during the estimating period to all parties whose work will be affected by these documents. They must also be prepared to answer questions and to provide clarification whenever requested. Ongoing communication and understanding between the general contractor and subcontractors is essential to effective bid assembly.

Final Review and Adjustments

Prior to completing a proposal, the estimating department must perform a detailed and comprehensive review of all work sheets, subcontractor and supplier bids, general conditions, supervision, overhead, profit, forms and documents, and all other related items. When corrections or adjustments are required, one final check should be made to ensure that they are correct in every respect.

Proposal Forms

It is essential to complete formal proposal forms, bid bonds, alternates, and other required documents with assiduous care. Alternates, when requested, are an integral part of the proposal, and the award of contracts is often based on the contractor's alternate prices. Therefore, they must be afforded the same degree of attention as the main body of the proposal. A final check to ensure that all forms and attached documents are completed as specified must be performed immediately prior to submitting a proposal.

If specified forms are neither furnished nor requested, the contractor's letter of proposal must be structured in an acceptable and appropriate manner and must include all pertinent information. In the case of a formal bid opening, that is, a specified time and place for the opening and reading of proposals, it is advisable, when time and conditions allow, for the contractor's representative to call the office for any last minute changes to the proposal. This particularly applies to the dollar amounts.

Cost Budgets

Once a general contracting firm has been awarded a contract, it is the estimating department's responsibility to prepare job cost budgets. This information is derived from the estimating work sheets, and includes all labor, material, equipment, rental equipment, general conditions, supervision, and all other expenses for work furnished directly by the general contractor. The composition of the budgets includes breakdowns for each work item as it relates to different areas of the structure, and specifies quantities and dollar value. The forms for the breakdowns should be organized to provide for entries of actual expenditures for each budget item.

Job cost budgets serve as a significant aid and guideline in assisting the field personnel with their production planning and cost control. They also assist the estimating department when updating the unit cost files for future projects. Prior to commencing each work item, the budget should be reviewed by the field supervisory staff, and then monitored and updated throughout construction. (See App. G.)

Problems

The three most common estimating problems a general contractor faces are mistakes in quantity estimates, incomplete or incorrect bids from subcontractors, and lack of sufficient time to prepare a proposal. Preventing these problems is the only effective solution.

Mistakes in quantity estimates are best prevented by persistent checking of quantities, unit prices, and mathematical equations and extensions. This checking is most beneficial when performed by one who did not do the original takeoffs. If this is not practical, the personnel that did the estimates must, in a systematic manner, check each item as the work progresses. Forgotten zeroes, misplaced decimal points, and improper formulas can have hazardous results. Check your estimates. It's important.

A prudent method of preventing mistakes in subcontractors' bids is for the contractor to insist that their bids be submitted in detail, allowing ample time for checking prior to the due date of the proposal. Also, communicating with the subcontractors at regular intervals during the bidding process is helpful in preventing mistakes and inadequacies in their bids.

The inability to complete a proposal due to lack of estimating time, can be avoided by estimating personnel's willingness to work even longer and more diligently than customary. In this event, additional temporary estimators should be employed to assist the estimating staff in the performance of their more mundane duties.

Chapter

11

Value Engineering Changes

Value engineering changes are changes to the plans and specifications in order to reduce the construction cost of a project. There are three circumstances under which these changes are most prevalent: during contract negotiations between owners and contractors; when the low competitive bids exceed the construction cost budget; and when requested by an owner after a contract has been awarded. Value engineering changes are distinguished from alternates, in that they may be proposed by a general contractor either prior or subsequent to being awarded a contract, whereas alternates pertain to items—specified by the design consultants—that are included in the specifications and bidding documents. Also, alternates often are for the purpose of providing owners and consultants with the opportunity to review materials, equipment, or systems that are more costly than those included in the base bid.

The most beneficial and cost saving value engineering changes become evident when the contractor works with the architects and engineers from the early design stages and throughout completion of the working drawings and specifications. This does not mean to imply that propitious results cannot be realized when working from the completed plans and specifications. But this method usually does not allow changes to be as extensive or cost effective. Also, extreme care must be exercised to avoid conflicts between various scopes of work when proposing changes from completed plans and specifications.

Due to the many contracts our company was awarded as a result of our proposed value engineering changes, it became common procedure for us to offer this service to owners and design consultants on projects we were bidding or negotiating. I recall two particular projects that exemplify the benefits of value engineering changes to a general contractor. The first project involved a large high-rise condominium building for which our company was one of the two low bidders. Arduous and ongoing negotiations ensued between the owner, our company, and our competitor. After approximately three weeks of back and forth negotiations, the owner was still undecided as to his choice of a contractor. Realizing that our competitor was offering approximately the same concessions and price as our company, we requested one additional week to review our proposal. Since we knew that to further reduce our bid price was not feasible, we decided to spend the time and effort required to prepare a list of proposed value engineering changes.

The owner's astonishment was evident when we presented him with proposed changes that reflected savings of several hundred thousand dollars. Based on our proposed changes, and subject to approval by the design consultants, the owner agreed to award the contract to our company. Most of our proposed changes and their related cost reductions were approved by the consultants, and we entered into a contract with the owner.

The circumstances surrounding the second contract were different in that the owner was a national corporation and the project was a sophisticated commercial building. In this instance, our bid was substantially lower than the next bidder. But the price exceeded the owner's budget, and, therefore, our bid could not be approved. We were requested by the architect to offer value engineering changes that would reduce the cost to the budget amount, thus enabling the project to proceed. A reasonable time frame was afforded us for this purpose, and we proceeded to prepare our proposed changes.

Due to the complexity of the design and the limited flexibility of the specifications, extraordinary resourcefulness had to be administered to successfully perform this work with results that would be acceptable to the owner and architect. This was accomplished, and the contract proved to be profitable for our company. Most gratifyingly, the owners, as a result of their satisfaction with the changes we submitted, and our performance, awarded a contract to us on a negotiated fee basis for one of their future projects. This would not have happened without the effective and productive application of value engineering changes. When the opportunity presents itself, and under favorable circumstances, general contractors should propose effective and meticulously prepared value engineering changes.

Value engineering changes often can produce acceptable cost savings related to every phase of a project. The range of items is vast, and may include every scope of work from structural systems to landscaping. When preparing proposed changes, a comprehensive review and analysis should be afforded each sheet of the plans and every section of the specifications. This work must be accomplished by experienced personnel who have total awareness and understanding of the project and are able to recognize practical changes that will result in cost savings.

Requesting suggested changes from subcontractors is one of the most useful means a general contractor can employ when preparing proposed value engineering changes to a project. Knowledgeable and experienced subcontractors can offer valuable suggestions and furnish the contractor with prices and other required information. It is the general contractor's responsibility to review carefully changes submitted by subcontractors, to be sure that they are in harmony with the design of the project and reflect price reductions consistent with their value.

Presentation of proposed value engineering changes must be in the proper form, with complete descriptions and cost breakdowns for each item. Their approval is often dependent on the contractor's ability to substantiate that the proposed changes will not adversely affect the project in its value, maintenance cost, or basic concepts and integrity as incorporated in the plans and specifications. This is particularly important related to the structural, mechanical, electrical, and other "behind the wall" items. Justification of the price reductions is always necessary.

There are several common problems that a general contractor faces concerning value engineering changes. During the process of preparing proposed changes, the contractor should consider these problems and the most effective means to prevent them.

Conflict of Work Items

General contractors are plagued by the problem that their estimators or subcontractors do not always recognize when a change to one work item will affect other work items. Changing the depth of recessed medicine cabinets from three inches to four inches necessitates changing the wall thickness. A simple example, but suppose that there are hundreds of medicine cabinets to be installed. Will substituting ceiling-mounted light fixtures for less expensive wall-mounted fixtures necessitate a more expensive routing system for the mechanical piping in the walls? The examples could fill volumes. These types of problems are magnified when value engineering changes are prepared from completed plans and specifications. Shop drawings and revised details on the working drawings will avoid conflicts in the field, but when conflicts due to changes are not considered and included in the pricing of proposed changes, additional cost will be incurred.

These problems can be avoided by thorough review and understanding of the proposed change items and how they relate to the balance of the project. This review is of a technical and detailed nature, and must be performed by personnel who are experts in this type of work. Nothing should be taken for granted when estimating proposed changes. The first question general contractors and subcontractors should ask themselves when pricing changes is, what work items, besides those that are obvious, will affect our scope of work? Prior to submitting proposed value engineering changes to an owner, the general contractor's staff should make the same final review, concerning conflicts of work, as performed when preparing the original job cost estimate.

Records and Notices

Failure to record approved value engineering changes and to notify all interested parties is an inexcusable problem that does, however, occur. The most efficient method of recording changes is to incorporate them in the working drawings and specifications. When this is not practical, addenda, detail sheets, and other written documents must suffice. When general contractors deem it unnecessary to notify subcontractors or suppliers not specifically involved in the changes, or are neglectful in advising those parties that are directly involved, unfavorable results will be evident.

Avoid these problems by including bold notes and details on duplicate copies of the working drawings, as they apply to the affected work items. Transmit these and other related documents to the field supervisory staff and all subcontractors immediately after receiving approval from the owner.

Design Consultants

Negative or unfavorable attitudes and unyielding positions displayed by design consultants impede the general contractor's efforts when proposing value engineering changes. While most consultants are receptive to changes that ameliorate or effect cost savings related to their plans and specifications, there are those who resent a contractor suggesting changes to their documents. The general contractor must recognize that these professionals believe that their own work reflects the best and most prudent methods, material, and equipment when attempting to alleviate these problems. A contractor should use tact and diplomacy when offering proposed value engineering changes, and project an attitude of cooperation and teamwork. Most important, the changes offered should not be frivolous or serve to compromise the concept or quality of the project. The contractor must be capable of qualifying the value of the proposed changes and of furnishing technical and other support data to substantiate their significance.

It is often helpful to conduct a preliminary review of the proposed changes with the consultants, prior to presentation to the owner. This often abates resistance to contractor proposed changes by the consultants and results in a more harmonious attitude by all concerned parties. Finally, the general contractor must elucidate all proposed value engineering changes astutely and professionally.

When offering proposed value engineering changes, a general contractor assumes responsibility to the owner, design consultants, and all involved firms and individuals. This responsibility must be honored with good intent and purpose. If these guidelines are not followed, value engineering changes can have an adverse affect on the reputation and successful development of a general contracting company.

Chapter 12

Schedules

Experienced general contractors recognize the significant worth of a job progress schedule. An effective schedule will provide invaluable assistance and information throughout a construction project. A job progress schedule functions primarily as a guideline that defines starting dates, allowable working days, and completion dates for each work item. However, a good schedule will furnish additional, equally significant information. Checking payment requisitions, projecting cash flow, and evaluating subcontractor performance give further evidence of a progress schedule's value. In addition, these schedules provide the information from which adjustments to cost estimates for future projects are determined.

The structure and composition of job progress schedules can range from a simple, hand-drafted, network bar graph to a computer-generated schedule that reflects each work item as it relates to each area of a project. The appropriateness of a schedule is determined by the requirements of the contract documents, the type of project involved, and the type of schedule with which the contractor has experienced the best results.

Effective job progress schedules must be in direct relationship to constructing a building in the field. This is best accomplished when schedules are developed by personnel who have field experience. In this regard, essential input can be furnished by the contractor's project managers and superintendents. Including all pertinent work items, assigning realistic time to complete each item, and scheduling the start of each item with proper lead time are the necessary ingredients to producing an effective job progress schedule. Lead time is established by relating the scheduled start of each event to its dependence on a percentage of completion of previous work. A comprehensive job progress schedule will incorporate the entire duration of a construction project, starting with move-in day—when construction starts—and ending with final punch list work. It will schedule every event requiring specific starting and completion dates.

Prior to preparing a job schedule, meetings should be conducted between the general contractor's staff and all major subcontractors. These meetings are for the purpose of allowing each subcontractor to establish the work days required to complete their respective work items on each floor and area. The starting dates of their work dependent on purchasing and receiving material and equipment are determined. It is the responsibility of the general contractor's staff to monitor these meetings to ensure that the subcontractors will be able to conform to the contractor's allowable time frames and to work out agreeable adjustments when deemed necessary. The meetings should be restricted to small groups of subcontractors whose work is closely interrelated. While this entails several meetings, it minimizes confusion and produces more beneficial results.

Upon completion of the preschedule meetings, all of the diverse information is reviewed by contractor staff, and documented. This includes work to be performed directly by the general contractor. At this point, development of the schedule is ready to commence. The first step is to prepare a network schedule that indicates the sequence of starting dates and allowable work days for each work item to be included in the schedule. From the network, a final schedule which indicates each event, on a floor by floor and area by area basis, is constructed. Included are the starting and completion dates, allowable work days, lead times, and connection to the preceding and subsequent work items.

Various ways to reproduce this information include lines, circles, codes, and other graphics. Alternatively, the schedule can be reduced to a bar graph that is developed from the multiple components but is less sophisticated in its structure. In either event, a means of tracking progress and updating must be provided for. Schedules that include the graphics of all the components are more illustrative and impressive than a bar graph. The bar graph is easier to read and more readily understood by field personnel, and, therefore, is more commonly used on the job site. The final schedule includes notes and graphics that define all codes, areas of the project, abbreviations, and other pertinent data that requires clarification. A computerized schedule will produce a printout that can be used instead of, or in conjunction with, the graphic schedule.

There are general contractors who structure their progress schedules so as to reflect a perfectly timed construction project. They employ the theory that the schedule can always be adjusted to conform to actual job progress. I believe more pragmatic results will be realized when this philosophy is reversed, that is, job progress should be adjusted to conform to the schedule. Since there are always unavoidable delays inherent to a construction job, reasonable time should be built into the schedule to compensate for these delays.

There are several available computer programs designed to generate construction schedules. But scheduling is so important and serves such a multitude of diverse purposes that I recommend that a general contractor invest the time and money necessary to develop a customized program compatible to the company's scheduling requirements. There are many firms and individuals experienced and trained in the technicalities required to develop customized scheduling programs. The quality and value of these programs depend upon accurate information received from the contractor's personnel.

The most common problems related to job progress schedules are described in the following paragraphs. Measures taken to avoid and effectively solve these problems are reasonably simple and uncomplicated if put into effect as soon as the problems become evident. This necessitates daily monitoring by the contractor's field supervisory staff, which should be expected under any circumstances.

Belief in Schedule

Too often, the subcontractor's and general contractor's staffs fail to give the schedule the credence it deserves. Like all phases of construction, progress schedules are not perfect, and when early recognition of weaknesses or errors in their structure become evident, it is not uncommon to hear "I told you so." This attitude can produce a domino affect, and must be avoided or corrected in the early stages of construction. If schedules are correct in their context, they must be adhered to. If this rule is not followed, the project will, more than likely, fall behind schedule.

Early correction of any errors, and advising the complaining parties that the corrections have been accomplished, will avoid repetition of this problem. In cases where doubt and lack of belief in the schedules is unjustified, it is incumbent upon the contractor to prove the schedule's worth, and to ensure that it is adhered to.

Change Order Work

Additional work or changes to existing work items can adversely affect a job progress schedule to the extent that the schedule's effectiveness is appreciably diminished. When change orders are approved, immediate meetings should ensue with all involved subcontractors in order to determine necessary schedule revisions due to the changes.

This procedure will help maintain balance and an even flow to the job progress, and avoid disruption to the scheduled work items.

Updating

Job progress, change orders, and payment requisitions are the primary reasons to update schedules. This process is time-consuming and costly. It should be done only when it is required. When specific work items fall behind schedule, updating obviously is required. However, production time may be picked up, thereby canceling the need for updating. Schedules used to measure completion of work for payment requisitions can prove inaccurate unless updated immediately prior to submitting requisitions. This, however, may not allow enough time to prepare the requisitions.

Avoiding the problem pertaining to work items that fall behind schedule is achieved when float time for noncritical work items is reflected in the schedule. Float time means the allowable days beyond the scheduled date to complete a work item, before there are delays to connecting work items. Those items that are critical from inception have no allowable float time. The problem of payment requisitions is avoided by updating the schedule five working days prior to the due date of each requisition, and including the work performed in the five-day periods in each subsequent requisition.

Chapter

13

Field Supervision

A general contractor will find no specific equivalent in value to proficient field supervision on a construction project. Assuming that the contract price and all other conditions are favorable, excellent field supervision is the dominant means of ensuring a contractor's success and profit. Conversely, inferior supervision will nullify profit potential and otherwise damage the general contractor's business.

Supervisory personnel assigned to a project must excel in the performance of their multiple responsibilities. Experience, fortitude, dedication, resourcefulness, and construction acumen are prerequisites for competent and productive field supervisory personnel. These qualities are most evident in people who have come up through the ranks. During this process, they have been exposed to the many diverse components related to construction. Therefore, they have a more comprehensive and personal sense of supervisory responsibility. As all contractors realize, but sometimes fail to emphasize, construction projects are built in the field and not in the office. It is the field supervisory staff that implements and follows through on all events and procedures, from commencement to completion. A general contractor who believes in these sound and valid principles will find no alternative to employing the most qualified and experienced field supervisory personnel available.

A proficient field supervisory staff is capable of relieving management from mundane responsibilities and problems in the field. Once construction of a project has commenced, that project should metaphorically be removed from the main office to the job site. Ideally, this principle applies to all phases and responsibilities of construction, short of top management decisions. Obviously, for this mode of operations to be successful, field supervisory personnel must have executive experience and characteristics, in addition to the aforesaid qualifications. Employing this policy allows management to direct their energies and efforts to seeking and acquiring new construction contracts, thus enabling the company to grow and prosper.

One of the more frustrating circumstances a general contractor faces is the inability to pursue a construction contract due to the lack of available, qualified, field supervisory personnel. Positive and aggressive characteristics indicate a successful general contractor. Yet, as enunciated earlier in this chapter, there are no alternatives for well-balanced and excellent field supervision. An experienced contractor is aware of the inordinate risk in undertaking the responsibilities of a construction project without qualified field supervision.

This dilemma amplifies the advantages of implementing job site training for potential field supervisory personnel. Careful monitoring and teaching of the people in training will expand and strengthen field supervision and provide personnel trained and disciplined in the operational procedures of their company. A result is the enhancement of a general contractor's opportunities to pursue additional work, confident that proper execution will be accomplished in the field.

Field supervision is most productive when the proper number of personnel is assigned to a project. This practice prevents overtaxing staff, duplicating responsibilities, and personnel conflict—any of which contributes to inefficient supervision. Supervision should be fine-tuned regarding division of responsibilities; that is, the experience and expertise of each person should be considered when delegating specific tasks. When undertaking construction of a building that is unique in design and intended use, as distinguished from projects familiar to a general contractor, project manager and superintendent selection must depend on their experience with similar projects.

Some years ago, we were awarded the contract to build a marine stadium. Although the dollar amount of the contract was not particularly large, the design and intended use of the structure was entirely different from any project previously built by our company. To this job, we assigned one of our experienced project superintendents, who had successfully supervised other complicated construction projects for us. Shortly after job commencement, cost overruns and delays became evident. Causes of these problems did not appear in the review of our cost sheets and estimates or in other investigative procedures. Failing to determine the problems through other means, we directed our attention to field supervision.

Verification that the problems were being caused by inept field supervision was quickly established. Although conditions improved to some extent by changing the project superintendent, our original profit projections were never realized. The problems encountered originated with executive management's failure to recognize the need of specialized field supervision. Had we been aware of this requirement, we would have realized that no one on our staff was qualified to supervise this type of project. Unless new personnel, having experience in this type of work, was available for employment, a prudent decision would have been not to bid the project. This experience taught us a costly, but significant, lesson. We never again endeavored to bid or construct a project with characteristics completely inconsistent with the experience of available field supervisory staff.

Notwithstanding the specific responsibilities assigned to individual supervisory personnel, the entire staff must be able to perform each other's duties. This is essential in the case of absenteeism, or while waiting for personnel replacements. It is imperative that the supervisory staff maintain ongoing communication to ensure their awareness of the progress, problems, and other items germane to an entire construction project. As previously stated, all activities of a construction project are dependent upon and connected to each other. Consequently, all scopes of field supervision must be closely coordinated to produce and maintain efficient, smooth-flowing construction operations.

In lieu of a graphically prepared organization chart, the following paragraphs identify field supervisory personnel and briefly describe their responsibilities. I believe this method is both more conclusive and more descriptive.

Project Manager

Accomplished project managers have that penetrating quality of leadership that evolves from technical knowledge and executive experience. They are the overseers and referees responsible for establishing and maintaining job balance and decorum. They settle disputes with owners, consultants, and subcontractors in a conciliatory manner. Project managers survey all construction activities but concentrate on the following aspects of field office and project management fundamentals.

Owners and owner agents

Project managers provide clarifications, details, and solutions to job problems that require owners' answers and direction; mitigate owners' construction-related concerns; perpetuate a cooperative attitude between the owner and contractor; conduct owner-contractor meetings; and maintain communications with owners and their agents. These are the five most prevalent owner-related responsibilities of a project manager. However, in all instances, the project manager is the general contractor's principal connection to the owner throughout construction.

Shop drawings, support data, and samples

The project manager prepares submittal data logs; reviews subcontractor's and supplier's shop drawings, support information, and samples; forwards these to owners' agents; distributes approved copies; and enters each transaction into the logs. The project manager is responsible for having these activities performed within the time frame set forth on the progress schedule and, in no event, to allow these items to delay job progress.

Progress schedules

The project manager makes daily reviews of work performed and tracks job progress; updates these as required; and distributes original and updated schedules to subcontractors and owners. The project manager is expected to make any changes that will improve the project schedules and to assist contractor and subcontractor personnel to understand and comply with the schedules.

Change orders

The project manager administrates all work pertaining to change orders. This includes preparing cost estimates, submitting estimates to owner for approval, and reviewing executed change orders for accuracy in all respects. Further responsibilities subsequent to approval include distributing revised plans, details, and documents to all interested parties, and revising the job progress schedule accordingly.

Budgets and records

The preparation and administration of job logs and records, and the review and filing of inspection reports, test results, minutes of meetings, and all other reports and records are the responsibility of the project manager. Reviewing cost budgets with the field supervisory staff and making adjustments when necessary are also included in this scope of work.

Requisitions for payment

This extremely important project manager responsibility includes updating the job progress schedule in time to prepare the requisitions; reviewing and recording all general contractor's direct job expenses;

confirming or adjusting subcontractor's request for payments; and assembling, and making a final review of, this data and other related information.

Project Superintendent

Enumerating the qualifications of project superintendents is best accomplished by stating that these people must be capable of performing all of the tasks assigned to the entire field supervisory staff. Experienced and competent project superintendents are the anchor people who provide fundamental stability to a construction project. These people are depended on to furnish direction and technical knowledge to their assistants and the subcontractor's field personnel.

Assistants' responsibilities

Project superintendents divide and delegate responsibilities to their assistants. This must be accomplished with awareness of each assistant's experience and proficiency as it relates to each scope of work. To realize the most beneficial use of all assistants, persistent checking of their work and daily communication is essential.

Coordination with project manager

Much of the information used by project managers, in the fulfillment of their responsibilities, is provided by the project superintendent. Therefore, the superintendent must work in close concert with the project manager. These parties work as a team, always aware of their common goals.

Subcontractor work

Since 85 to 90 percent of the work items in a construction project are performed by subcontractors, it is obvious that the prime responsibility of a project superintendent is the satisfactory performance and completion of the subcontractor's work. The project superintendent is responsible for the progress, quality, coordination, and all other items pertinent to the work included in the subcontractor agreements.

A common tendency of many subcontractors is to become remiss in their efforts during completion of the last 10 to 15 percent of their work, including punch list items. It is incumbent upon the project superintendent to require their reinforced efforts when this weakness becomes apparent. When subcontractors fail to complete their work, the project superintendent must exercise prudent judgment in authorizing this work to be performed by others.

Work performed directly by the general contractor

Project superintendent responsibilities include close surveillance of all labor, material, rental equipment, and other expenses for work performed directly by the general contractor. Despite the cost budgets pertaining to these items, savings should be realized when possible, and cost overruns kept to a minimum. This can only be achieved by exercising the astute judgment of a competent project superintendent.

The project superintendent must demand and receive the same quality, progress, cooperation, and efficiency related to the general contractor's work items as is expected of the subcontractors.

Job morale

Establishing and maintaining good job morale is essential to a successful construction job. The demeanor displayed by the project superintendent has significant influence on the quality of personnel morale prevalent on the job. Earned respect, strong leadership, and general construction acumen are attributes that must be evident in the project superintendent, if favorable job morale is to result. Motivating, without dictating, serves to accomplish this end.

Structural Superintendents

Structural superintendents are accountable for all work pertaining to the structural system of a project, from layout to completion. This particularly applies to six categories of structural supervision: quality control; adherence to progress schedules; coordination of work items; problem recognition and solution; the most beneficial use of labor, material, and equipment; and record and report maintenance.

Close coordination with nonstructural work items is necessary to avoid conflicts and mistakes. Effective structural supervision requires constant on-site presence of the structural supervisory staff and personnel experienced and dedicated to their work.

Architectural Superintendents

Essentially, the same responsibilities apply to architectural superintendents as to structural superintendents. Work under their surveillance includes all items not directly related to the structural, mechanical, plumbing, and electrical systems.

Mechanical, Plumbing, Electrical Superintendents

Specialized field supervision for these items is furnished by the general contractor for projects employing unique and complex installations. Otherwise, the structural and architectural superintendents, along with the project superintendent, provide the necessary supervision related to these installations. Of critical importance to such scopes of work is the checking material, equipment, and installation methods, prior to covering up, to ensure that they are in conformance with the plans and specifications.

Interior Design Superintendents

If interior design work for public areas is complicated and extensive, specialized field supervisory personnel is required. These superintendents must be well trained, experienced, and familiar with this type of work.

The ability to clearly understand the intent of the design consultants and obtain information and details not included in the interior design drawings and specifications is of paramount importance. This type of work usually progresses at a slower pace than other work items; therefore, the superintendent must establish and enforce the earliest practical commencement date. Making sure that the work is protected during construction and uncompromising quality control are important functions of this superintendent.

Punch List Superintendents

This personnel is responsible for inspecting all finished work, and using sagacious judgment in determining what items require repairs, adjustments, corrections, or replacements. They document this information and distribute it to the responsible subcontractors. While the punch list work is being performed, all procedures, starting with inspections, are repeated until satisfactory completion is accomplished.

Labor Superintendents

This supervision relates to miscellaneous labor items when performed directly by the general contractor. The duties included are: distribution of materials, job cleanup, safety work, maintenance of tools and equipment, and general-condition items not specifically mentioned. The labor superintendent's responsibility for cost effective procedures, quality control, timely completion of work, and accurate records and reports is of equal importance to that of all other field supervisory staff members.

Field Office Support Staff

Clerk of the works, bookkeepers, and secretaries are not considered supervisory staff. However, their jobs are directly related to job management, and they perform under the direction of the project manager and project superintendent. These are trained personnel, experienced in the peculiarities of the construction business.

Clerk of the works

This individual's primary duties include ordering and checking material, filing plans and construction related documents, controlling keys, approving purchases of small tools and minor repairs, releasing and checking off stored materials and equipment, and furnishing general office assistance to field supervisory personnel.

Field office bookkeeper

When bookkeeping is performed in the main office, the field office bookkeeper reviews and checks all support information prior to forwarding it to the main office, and maintains duplicate files in the field office. The bookkeeper acts as the payroll timekeeper and assists the clerk of the works and the field secretary. If bookkeeping is performed in the field office, the work enumerated in Chap. 5 ("Bookkeeping") applies.

Field office secretary

This work includes receptionist and general clerical duties.

A general contractor's field supervisory staff, as demonstrated in this chapter, is comprehensive; the entire staff would be utilized only on a large construction project. Obviously, the number of personnel and their assigned responsibilities vary from one project to another. But the principle concerning qualified and experienced supervisory staff as elucidated in the foregoing paragraphs remains constant, and applies to any type or size of construction project.

Chapter

14

Preconstruction Meetings

Normally, a minimum of four preconstruction meetings are required to set the tone, provide in-depth understanding, and establish procedures for construction of a new project. A detailed agenda, carefully prepared in advance of the meetings, is used as a guideline to review each specific item. Although an informal, relaxed attitude should prevail at the meetings, all reviews and discussions must be goal orientated and treated with accuracy and professionalism.

First Meeting

The general contractors' management team, field supervisory staff, and estimating personnel must be in attendance at this meeting. The meeting should start with a general review of the project, with special attention to any unusual circumstances and conditions. Once the review has been accomplished, the following items are discussed in detail.

Plans, specifications, and addenda

Upon detailed examination of these documents, ensuing questions, answers, comments, and suggestions take place until all personnel have a complete understanding of, and are in agreement with, the proposed methods and procedures. All pertinent items resolved at the meeting are recorded and distributed to the staff members. The required number of approved working plans, specifications, and related documents are numbered and provided to the field supervisory staff for distribution and field use.

Field supervision responsibilities

The specific responsibilities of the field supervisory staff is established. To provide a complete understanding of their individual responsibilities, a detailed review is conducted of all work items assigned to each staff member. The project superintendent is the primary person making these decisions.

Work performed by general contractor

Those work items to be performed directly by the general contractor are agreed upon. The most efficient and cost effective use of labor, material, and equipment is determined for each work item. This information is recorded and distributed to the field supervisory staff.

Forms, Records, and Logs

Instructions are given to the field personnel concerning the manner in which these printed forms are to be completed. They include payrolls, change orders, weather reports, listings of subcontractor personnel on each job, records of meetings, shop drawings, budgets, back charges, and requisitions.

Subcontractors and suppliers

The individual subcontractor agreements are reviewed in detail, with particular reference to any peculiarities as to scopes of work or conditions included in each agreement. Items and conditions agreed upon between the general contractor and subcontractors not specifically included in the agreements are reviewed and documented. The field supervisory staff is made aware of all suppliers and rental equipment companies to be used on the project. A list including their names, phone numbers, price lists, and other pertinent information is furnished to the staff.

Job progress schedules

This meeting allows the contractor's staff to collectively examine and fine-tune the schedule. With the field supervisory staff in attendance, practical revisions often occur due to the input of personnel experienced in specific scopes of work. Computerized schedules can easily be revised to incorporate these last-minute changes, prior to distribution to the subcontractors, owners, and other interested parties.

Cost budgets

This meeting is the most opportune time to examine and discuss the budgets in detail. Since the budgets are prepared from the job cost estimates, it is the estimating department's responsibility to explain and justify the budget amounts for the individual work items.

Review of the budgets provides a clear understanding of the labor, material, and equipment allowed for the various work items performed directly by the general contractor. Using this information, the field superintendents can better prepare and organize their work prior to starting construction.

Commencement of construction

These discussions include establishing the start-up date, location of field office and contractor's signs, moving of equipment on site, and general methods and procedures applicable to commencement of construction. This portion of the meeting may be expanded to discuss company morale and to instill feelings of confidence and optimism in the entire staff.

Second Meeting

The attending parties should include principals and representatives of the general contractor, the owner, architects, and engineers. As the meeting begins, the general contractor should exert efforts to establish a comfortable and friendly atmosphere. The meeting will be more productive when a spirit of teamwork and cooperation among all parties is evident.

Introductions

As introductions take place, each person's assigned responsibilities should be clearly identified. When reviews and discussions concerning all pertinent items have been concluded, independent meetings between the design consultants and the contractor's field supervisory personnel should take place. This procedure provides an excellent means for the consultants and superintendents to express their concerns and to resolve anticipated problems.

Change orders, shop drawings, and requisitions

Agreed-upon methods and procedures regarding these aspects of the project are established at this meeting. Sample forms are reviewed for acceptance and are revised if necessary.

Owner-furnished items

The general contractor should specify, in documented form, the required delivery and installation dates for owner-furnished items that will be performed during construction. Furnishing this information to the owner at the preconstruction meeting helps prevent unfavorable results at later dates. When possible, samples, details, and installation drawings of the owner-furnished items should be furnished to the contractor at this meeting.

Owner's special requests

This meeting provides the owner an excellent opportunity to make special requests. These requests may pertain to owner's use of building prior to completion of the project, reports and information additional to those specified, desired access and storage areas, and any other items of concern to the owner that are not specified in the bidding or contract documents. Also, the design consultants should take advantage of the meeting to request the contractor's compliance with procedures, peculiar to their firms, that are not identified in their specifications.

Job progress schedule

A review of the schedule is accomplished at this meeting. When requested by the owner, it is incumbent upon the general contractor to explain the details and intricacies of the schedule. Realizing that a schedule is an invaluable tool to a construction project, it is important for owners and their agents clearly to understand the schedule's structure and purpose. Although this often requires patience and explanations of a technical nature by the contractor, the time and energy expended is justified.

Value engineering changes

Since the owners and their agents are collectively in attendance at this meeting, it is the most convenient time to conduct a final review of all previously approved value engineering changes. Resolving any last minute questionable items is more effectively performed in the presence of all interested parties.

Third Meeting

This meeting includes staff members of the general contractor and prime subcontractors. Once all items on the agenda have been briefly reviewed and discussed, separate meetings should take place with individual subcontractors to analyze each portion of their respective scopes of work. This method requires one large meeting followed by several smaller meetings, but it is the most direct and efficient means to cover all important items.

Subcontract agreements

Review of these agreements is for the primary purpose of ensuring that everyone has a clear understanding of the work to be performed by each subcontractor. The list of items included, methods and systems, obligations and responsibilities are all closely examined. By exchanging views, issues concerning work not specifically identified in the bidding or contract documents, but expected to be performed by the subcontractors, are resolved and documented.

Frequently, the general contractor's and subcontractor's principals and upper management personnel have a more comprehensive understanding of the subcontract agreements than do their field supervisory staffs. Field personnel tend to consider the agreements as a means of documenting the work to be performed at an agreed-upon price, but fail to fully understand the legal ramifications and other important items included in the agreements. This meeting affords the opportunity for all parties to gain an understanding of the complete contents and authority of the agreements.

Clarifications

Prior to the meeting, the subcontractors, in brief written form, should advise the general contractor of any items requiring clarification. This request will enable the contractor, in advance of the meeting, to prepare answers to their questions and to provide the necessary clarifications. Such a procedure allows time for the contractor to fully understand the subcontractors' concerns and avoid delays when complying with their requests. Notes should be made of this portion of the meeting and distributed to all interested parties.

Job progress schedule

An extensive review of the schedule is performed to assist the subcontractors to thoroughly understand its structure, purpose, and importance. This meeting affords the subcontractors an opportunity to request schedule changes relating to their specific work items. The changes may apply to starting dates, time allowed, and the inclusion of specific items not previously considered.

Shop drawings and submittal data

Instructions regarding the preparation, necessary forms, required number of copies, and other related information essential for submitting subcontractor-furnished shop drawings and samples are reviewed. At this time the general contractor should reemphasize the importance of properly prepared data being delivered on, or before, the dates in the job progress schedule.

Forms

Subcontractor use of standardized forms is necessary for efficient processing of paperwork throughout construction. Samples of these forms are reviewed, explained, and distributed to the subcontractors. Necessary revisions to certain forms, due to unusual circumstances related to individual subcontractors, are made at the meeting. Progress reports, back charges, change order requests, and inspection reports are examples of such forms. A printed copy of instructions is attached to each set of forms.

Payment requisitions

This portion of the meeting starts with distribution of several documents: requisitions, release of liens, certification affidavits of work performed and amounts paid and due, and breakdown of work including dollar value. Once these documents have been explained in detail, the importance of their being accurately completed and furnished on time is emphasized.

The subcontractors are advised of the absolute requirement to submit their payment requisitions for work actually performed, and in compliance with the contract documents. The unfavorable consequences of requesting payments in excess of work completed is asserted by the general contractor in a positive and strict manner. (See Apps. A to C.)

Storage and parking areas

Designated areas for subcontractor-furnished materials, equipment, and personnel parking are specified. This portion of the meeting extends to instructions concerning on-site fabrication yards, location for equipment used to hoist and distribute materials, and locations for field offices and tool sheds.

Fourth Meeting

Attending this final preconstruction meeting are the same members of the general contractor's staff who were in attendance in the three previous meetings. There are three primary purposes for this meeting: to recapitulate the pertinent items and information reviewed in the previous meetings, to ensure that clarifications and changes resulting from the meetings are properly documented, and to reinforce a spirit of teamwork and gainful morale in all personnel. Additionally, this meeting affords an opportunity for staff members to express last-minute concerns, offer final suggestions, and prepare themselves mentally for the task ahead.

Proper advance preparation and effective execution of preconstruction meetings produce boundless advantages to a general contractor starting a new construction project. New methods, systems, and procedures are discovered, potential weaknesses and strengths are revealed, administration details are established, instructions are defined and explained, differences are resolved—these are all advantages. During the course of construction, a percentage of the items understood and agreed upon at each meeting will be neglected or forgotten. Periodic communication with those parties in attendance is essential to perpetuate the value of the original meetings. This requirement applies to the general contractor's staff, the owners, and the subcontractors.

Chapter 15

Cost Control Records, Field Production Files, and Requisitions

Cost Control Records

Breakdown cost sheets for labor, material, and equipment furnished directly by the contractor, and summary cost records for all subcontractor-performed work are included in the cost control records. These records are of equivalent value to controlling job costs as are plans and specifications to constructing a project. The attitude "we will do the best we can" is not acceptable. Instead, realistic budgets, judicially prepared, must be the authoritative documents used to guide the general contractor's staff throughout construction.

Cost control records provide the information that alerts us to cost overruns or savings and in determining if these variances are caused by inaccurate estimates, ineffective job management, or other contractor-controlled factors. The knowledge gained from these records tell the general contractor when it is necessary to adjust estimating formulas and to improve methods and procedures.

To realize beneficial results from cost control records requires frequent meetings between the general contractor's and subcontractor's field supervisory staffs. Reviewing job costs, discovering reasons for overruns and savings, and updating the cost records to reflect actual expenditures are accomplished at the meetings. With the aid of computers, the mechanics of updating and maintaining accurate costs records becomes simplified. Also, computer prepared and updated records command more attention from the field personnel. (See Apps. G and H.)

Field Production Files

There are three basic categories of construction project files: those described in Chap. 5 ("Bookkeeping"), legal documents pertaining to disputed items, and the field production files identified in the following paragraphs. Copies of the field production files are kept in both the main and field offices. These files must be not only readily accessible to the contractor's supervisory staff but also maintained and updated in an orderly manner throughout construction. Once a project is completed, the files should be clearly labeled and moved to the main office.

Project contract

This file includes three specific articles: the formal contract executed by the contractor and owner, a complete set of initialed contract documents identified in the contract, and executed change orders. Attached to the file is an updated recap record listing all approved change orders in their numerical order, and any other changes or agreements that affect the dollar value or time of completion, or in any other way revise the contract.

Change order work files

This file consists of correspondence between contractor, owner, and subcontractors, estimating sheets, subcontractor proposals, and other information concerning individual pending change orders. Once a change has been approved, the corresponding number indicated on the formal change order is assigned to this data. This file becomes a source of reference for updating cost budgets, submitting change orders to subcontractors, and contains all backup information and details related to each approved change order.

Estimating files

Included in these files are the worksheets, spreadsheets, recap sheets, subcontractor bids, and other support information from which the general contractor's original proposal is developed. Corrected and updated copies should be attached—in their chronological order of occurrence—to the original documents.

Subcontractor agreements

This file contains the original executed copies of all contracts between the general contractor and subcontractors. Any documents that change the price or completion or otherwise affect the essence of the contracts are attached to the agreements. Letters or notices with potential legal ramifications—concerning subcontractor failure to properly perform or other noncompliance items—are attached to the individual agreements and become a part of these files. These files should also include agreed-upon price lists and related information furnished by material suppliers and rental equipment firms.

General correspondence: Contractor-owner-consultants

This file contains correspondence of the many diverse items not specifically identified in this chapter. Questions and answers regarding miscellaneous information and details, disputed matters not of legal consequence, and explanations of methods and procedures are but a few of the items included in this file.

Often a letter, telegram, or other form of written communication makes reference to matters included in specific files. In this case, it is advisable to place copies of the written communication in both the specific and general correspondence files. This provides an efficient method of cross referencing and easy access to individual communications.

General correspondence: General contractor-subcontractors

This filing system requires individual files for each subcontractor. In principle, the same type of general correspondence as between contractor and owner is contained in these files, and the same method of cross referencing is employed. Due to the diversity of daily events, the volume of correspondence between the contractor and the subcontractors will substantially exceed that included in the contractor-owner files. In addition, these files include correspondence between the contractor and suppliers.

Job progress schedules

All correspondence, transmittals, and other documents between the contractor, owner, and subcontractors, that relate to job schedules are kept in this file. Included are copies of submittal letters or forms attached to original or updated schedules, notices to subcontractors or suppliers of their failure to perform in accordance with the schedule, and written schedule revisions that do not require updated schedules.

Inspection reports

This file contains three categories of documented field inspection reports, those performed by governmental agencies, architects, and engineers. These reports are not to be confused with the reports described under the following heading, but instead concern specific work items required by law and the contract documents to be inspected as the work progresses. The majority of these inspections are performed by the various building department inspectors and the structural, electrical, and mechanical engineers.

Architect and engineer job reports

This file pertains to all job-related reports prepared by the architects and engineers. The contents of these reports may refer to such specific items as job problems, quality control, or progress, or may be of a more general nature. Their purpose is to keep the owner aware of the job status as well as any unusual conditions or circumstances.

Test reports

This file contains test reports of materials and equipment required by the governmental agencies and the contract documents. This file often provides valuable and conclusive evidence used in arbitration and litigation. Test reports should be carefully reviewed by the contractor's staff prior to filing.

Job logs

Daily job logs, recapitulated weekly on standard forms, include five basic items: weather conditions, subcontractor personnel on the project, brief summary of work performed, meetings held, and unusual events. When deemed important, copies of a log containing important information concerning specific items should be attached to the documents in their corresponding files. Some examples are: loss of a workday due to inclement weather; an illegal strike, causing work stoppage. In the event these occurrences necessitate additional time in order to complete a project, copies of the logs that contain this information should be attached to the contract between the owner and general contractor.

Shop drawings and submittal data

Individual files are provided for each subcontractor and supplier furnishing shop drawings, samples, and other related submittal data. These files also include color and material selections. These are the most voluminous files in a contractor's filing system and are frequently accessed.

Minutes of meetings

Minutes of the meetings with owners, consultants, and subcontractors are kept in this file. These minutes are often referred to throughout construction and should be filed in chronological order. Whether they are prepared by the contractor, consultants, or other parties they are stored in this file.

Progress reports

Progress reports, in whatever graphic form the contractor chooses to use, are kept in this file. These reports often break down areas of work items in greater detail than does the job schedule. An example is that the schedule will show drywall board work on the fifth floor; the progress reports may break down this work on the fifth floor into two or more sections, with percentages of completion for each section. These reports also serve as an excellent means to check payment requisitions. Therefore, the progress reports are of significant value when evaluating job progress and dollar value, and deserve an accurate and orderly filing system.

Punch list data

Written communications between the general contractor, owner, and subcontractors including correspondence, inspection reports, detail lists, and all other documents pertaining to punch list work are a part of this file. The majority of the file contents will consist of correspondence, and breakdown lists of required work prepared by the general contractor.

Since this file usually becomes voluminous, especially on large projects, it is advisable at frequent intervals to consolidate the itemized work items remaining to be completed, and to place the previously prepared lists into an inactive file. Obviously, at this stage of construction, only the incomplete work requires intensive concentration, and this method of filing will save time when referring to incomplete work items.

Warranties, guarantees, as-built drawings

Due to the bulk of these documents, a large area must be provided for their files. However, these items require the simplest of all filing systems, as they are not received or filed until the end of the project.

The number of files and records enumerated in this chapter, in addition to the legal and bookkeeping files required for each construction project, emphasizes the need for an efficient filing system. Reference to project files, years after job completion, is common in a general contractor's office. Litigation long after project completion, a source of valuable information for future projects, and resource data used for teaching new staff members—all are valid reasons to afford the project files the attention commensurate with their importance.

Requisitions

Payment requisitions are the heartbeat of a general contractor's business. Without receiving payments promptly and in proper amount, a construction project dies. Strict attention to details and uncompromising accuracy must be the enforced measurement of standards a contractor employs when preparing requisitions. These same principles apply to the support data and affidavits accompanying the requisitions. The general contracting business is detail-oriented. Payment requisitions epitomize this fact.

Prior to submitting the first requisition for payment, the contractor should furnish the owner a sample requisition for approval. This affords the owner the opportunity to approve or make agreed-upon changes to the requisition form, break down work items including their respective dollar value, and any other related items included in the requisition. Included in this sample package should be the supplementary forms to be used for change orders, required affidavits, and all other documents that will accompany the requisitions throughout the project.

A general contractor should use forms accepted by the industry and familiar to owners, consultants, and lending institutions. Computer-generated requisitions have three distinct advantages: timesaving preparation, reduced risk of mathematical errors, and updating ease. Also, when using computerized requisitions, a contractor's standard form can be easily adapted to the specific type of forms required by the contract documents. (See App. E.)

Many projects require several parties to review a contractor's requisition. This usually includes owners, architects, engineers, and lending institutions. To ensure that ample time is allotted for these reviews, it is of paramount importance for the contractors to submit carefully prepared and checked requisitions on, or before, the dates set forth in the contract. Failure to do so will often result in late processing and payment of the requisitions.

Remembering that the purpose of this book is to provide useful information for the general contractor, I feel obliged to make the following comments concerning general contractors' payment requisitions. Unfortunately, it is not an uncommon practice for owner's representatives to reduce the payment amount of a contractor's requisition without proper justification. This is not meant to imply that agreed-upon adjustments are not frequently necessary. However, these types of adjustments must be distinguished from arbitrary or unjustified reductions in payments.

When faced with this frustrating and time-consuming situation, the contractor must, in an assertive and businesslike manner, insist that the requested amounts be promptly paid. Obviously, prior to making these demands for payment, the contractor must be able to substantiate that the requisitions are properly prepared, the requested amounts represent actual work completed, all bills and obligations have been satisfied, all forms and affidavits have been properly executed, and that the contractor has otherwise performed in accordance with the contract documents.

The same standards and principles regarding the general contractor's payment requisition apply to subcontractors. Dates for submittal, forms and affidavits to be used, breakdowns required, approved schedule of payments, and all other related items are established at the preconstruction meeting. It is the general contractor's obligation to inspect the work performed by each subcontractor prior to approving, adjusting, or disapproving the subcontractor payment requisition. This also applies to sums requested for materials and equipment stored on the site or at other agreed-upon locations.

The general contractor, after processing subcontractor requisitions, must make payments to the subcontractors promptly and in the approved amounts. Extend the same consideration concerning payment requisitions to your subcontractors as you expect from the owner. The general contractor who follows this principle will receive lower bids and better efforts than those contractors who take advantage of their subcontractors and use any excuse to delay making payments.

Chapter

16

Shop Drawings and Change Orders

Shop Drawings

For the purposes of this chapter, shop drawings are drawings, printed data, and samples that illustrate, describe, and provide details related to certain materials, equipment, and methods of installation. The project specifications identify the specific scopes of work and items for which such drawings must be provided. Architect's and engineer's approval of the information they contain is required prior to incorporating the work into a construction project. Shop drawings are, normally, prepared and submitted to the general contractor by the responsible subcontractors, equipment manufacturers, and suppliers. Once they are checked and processed in the contractor's office they are forwarded to the architects, engineers, or other designated owner's representatives for review.

Shop drawings are valuable means of providing information, details, and clarification to supplement the project plans and specifications. These drawings are depended on to assist the general contractor and subcontractors in the coordination and proper performance of their work. Contractors too often do not realize the value of shop drawings, and consider them to be a time-consuming and unnecessary requirement of the contract documents. This is a misapprehension. Drawings should be afforded the same degree of attention as are the plans and specifications.

It is not uncommon for design consultants to inadvertently approve shop drawings that contradict the details depicted in their plans. Contractors must be aware that this does not relieve them of their responsibility to perform the work in accordance with the contract documents. But when the plans and specifications regarding specific items do not include details or information, so that the consultants are completely dependent on the shop drawings to provide this data, it is incumbent upon the consultants to accept the responsibility of their approvals. When this principle is not adhered to, unfavorable circumstances may result, as the following events experienced by our company will illustrate.

Railings specified to be installed on the terraces of a building under construction were precast concrete. The architect's design and specifications were limited to dimensions, configurations, and material. On the plans, he included notes requesting that all details concerning the construction and installation of the railings be included in the shop drawings furnished by the manufacturer, and submitted to him for approval. Realizing that the architect was dependent on the shop drawings for full details and information, we thoroughly checked the drawings before forwarding them to his office. Upon completing our review of the reinforcing steel, grouting materials, installation methods, and all other related details we returned the drawings to the manufacturer for corrections and additional information. The second set of drawings was checked and approved by our office and submitted to the architect for his review and comments. He returned the drawings to us clearly stamped and initialed "approved as submitted." These approved drawings were returned to the manufacturer, and fabrication commenced.

The railings were installed on the first four floors before the architect determined that certain fabrication and installation details did not comply with his intent or requirements. He demanded that, at substantial expense to our company, we make the changes he desired. Facts to remember: the architect included no details or information in his plans and specifications; he requested this data be included in the shop drawings; he approved the shop drawings in their entirety. This was a definitive example of an architect's attempt to burden the contractor with the consequences of his own professional failings. This matter was resolved by our furnishing reports, prepared by independent consultants, which stated the fabrication and installation was proper and in accordance with all safety requirements and codes.

General contractors must be wary of architects and engineers who return shop drawings stamped "reviewed as submitted" with no other comments or notes. Design consultants that engage in this practice are shirking their responsibility, and reduce or eliminate the effectiveness and true purpose of shop drawings. I have heard certain consultants claim that it is not their responsibility to correct submittal data; their insurance company will not allow them to comment other than "reviewed as submitted"; it is a contractor's sole responsibility to submit shop drawings that, in all respects, conform with the contract documents. These statements are erroneous and almost completely defeat the advantages of preparing and submitting shop drawings.

Furthermore, if this position were honest and well intended, then why do project specifications, prepared by the design consultants, demand that shop drawings be submitted for their approval prior to fabrication or installation of those items requiring shop drawing submittal? Shop drawings are intended to confirm that proper construction methods, techniques, material, and equipment are utilized. The responsibility of meticulous review, constructive comments, and required corrections by the design consultants is of equivalent importance to the exactitude shop drawings should be prepared with. The appropriate notes or stamps applied by design consultants to shop drawings are "approved as submitted," "approved as noted," or "disapproved—resubmit." I strongly urge that contractors not accept shop drawings returned from a consultant stamped "reviewed as submitted," when no other comments or notes are included.

Logs

Prior to commencement of a project, the general contractor must prepare logs that reflect every transaction related to shop drawings. Forms are constructed to record the names of each firm submitting shop drawings, dates transmitted or received, and identification of items included in each submittal. This same information is recorded when the shop drawings are forwarded to the owner's agents, returned to the general contractor, and finally returned to the firms that originally prepared and submitted the shop drawings to the general contractor. In the case of resubmittal, the same entries are recorded in the log.

The logs serve the important function of providing the contractor with an organized and updated processing status of all shop drawings. This enables the contractor to follow up when subcontractors, suppliers, and design consultants are delinquent in submitting, returning, and processing the shop drawings. Consequently, the logs are frequently referred to and must be accurately maintained.

Submittal forms

General contractors, subcontractors, and suppliers should use forms of the same basic design for transmitting shop drawings. This is easily accomplished when the general contractor provides sample forms at the preconstruction meeting. The forms must have space for all appropriate information and include a section for remarks, approval status, dates, and signatures. These forms are attached to all submittal data and filed in the shop drawing files.

General contractor review

It is incumbent upon the general contractor to carefully review all shop drawings for completeness, accuracy, and compliance with plans and specifications, prior to forwarding them to the owner's agents. Corrections made by the contractor must be clearly noted on the drawings, and initialed by the person assigned to this task. When it is determined that the shop drawings have excessive errors, are incomplete, or are otherwise unsatisfactory, the contractor must return the drawings and request that they be resubmitted.

While it is preferable to have one member of the contractor's staff review all shop drawings, it is not always practical. Certain submittal data require structural, mechanical, electrical, or other specialized backgrounds. If the project manager, or other assigned staff member is not qualified in these areas, personnel with the necessary training should be employed for this purpose. Also, certain projects, due to their size or complexity, produce such voluminous quantities of shop drawings that it requires more than one person for review and processing. Should a contractor not have personnel available for this work, independent firms can be engaged to supplement the staff.

Timely processing and distribution

General contractors assume great risk when they perform work prior to receiving shop drawing approval from the design consultants. They are also exposed to great risk when they fail to complete a project on time. Therefore, it becomes evident that timely processing and distribution of shop drawings is essential. The most effective means to avoid delays with respect to shop drawings is for general contractors to frequently refer to the logs, and to provide persistent follow-up when necessary. The contractor cannot afford to hesitate in demanding that all involved parties prepare, transmit, review, and return submittal data within a reasonable time frame.

Change Orders

In the introductory chapters of this book, I spoke of the ecstasies and agonies of being a general contractor. In my opinion, change orders rank high in the agony department. I have often fantasized about constructing a building that required no change orders. And fantasy it is.

Of course, agony is not always the predominant emotion regarding change orders; they sometimes produce a reasonable profit to the contractor, provide needed clarifications, and generate a better product, but, more often than not, change orders cause delays in job production, create stressful relationships between owners, consultants, and contractors, and detour the energies of the contractors' field supervisory staff.

Estimated contractor profit concerning change order work is often not realized—certainly not to the extent many owners and owner agents may imagine. If the additional time required by the contractor's estimating department, clerical staff, field supervisory staff, and overhead expense could be accurately analyzed, a contractor would often find the anticipated profits disappointing, and capital losses not uncommon. Another conspicuous factor that contributes to the agonies of change orders is the potential for unforeseen delays to job production.

Notwithstanding the dismal characteristics I have attributed to change orders, they are a fact of life in a general contractor's operations, and must, therefore, be afforded the contractor's most intense vigilance. An attitude of indifference or "not my problem" displayed even to the sometime frivolous change order requests by owners does not exemplify an experienced general contractor. Errors and omissions in the plans and specifications, owner-desired changes, general contractor and subcontractor requests are all valid reasons for change orders.

Before enumerating specific items of importance pertaining to the processing of change orders, I emphatically assert that a general contractor should not perform change order work prior to receiving a formally executed change order. The only exceptions to this hard and fast rule may be work related due to life-threatening circumstances, potential property damage, or other emergencies. Even under these circumstances, when the contractor is entitled to compensation, all extra work performed should be immediately documented, and a change order request submitted without delay.

Pricing

Pricing change order work demands the same extensive review of the plans and specifications as elucidated in Chap. 11 ("Value Engineering Changes"). Serious conflicts can result to related, or seemingly nonrelated, work items when a contractor does not carefully check to determine if the work included in a change order will affect other scopes of work. It is the contractor's responsibility, when pricing change orders, to include price adjustments for all work influenced by the specific change items, and, when justified, to request additional time to complete the project. Once a change order has been executed by the owner, the contractor can not expect and, usually, will not receive compensation for the items that were overlooked in the original change order request submitted to the owner.

In preparing a cost estimate for change orders, the same type of recapitulation and spreadsheet should be employed as described in Chap. 10 ("Estimating"). Summary types of proposals submitted to owners should be avoided, as change order requests are more effective when they include cost breakdowns and all related information.

Follow-up

Frequently, advice of potential change order work is furnished to the general contractor in verbal, or in informal note, form. Persistent follow-up by contractors is often necessary to obtain required details and information, prior to review and pricing, and is essential when owners are delinquent in advising contractors if their change order requests are approved or disapproved. This same type of follow-up must be employed when change orders have been approved but the formally executed change order has not been forwarded to the contractor within a reasonable time.

When subcontractors and suppliers are remiss in furnishing the general contractor with pricing and other change order related information, follow-up by the contractor is necessary. It is interesting to note that certain subcontractors desiring change orders will be very prompt when submitting prices, but will procrastinate in furnishing prices and information when the change order is requested by others, and does not reflect financial gain for their firm. This is an item for a general contractor to consider when awarding subcontractor agreements.

Processing and updating

Once a change order has been formally approved, a general contractor must immediately notify all involved subcontractors, and other interested parties. I must confess our own company's failure to observe this rule on one occasion. It was on this occasion that our neglect resulted in extensive corrections to an important scope of work. Obviously, we paid for our negligence. It is another clear example of the small margin of error allowed a general contractor before unfavorable consequences become evident.

It is equally important to expedite the furnishing of all processed, approved change order paperwork to the contractor's field supervisory personnel. This affords them the opportunity to become familiar with the changes, resolve any field problems, and have the change order work commenced at the earliest possible time. In short, processing change order paperwork must not be shoved aside as something to be done tomorrow. The same principle should be administered to this important ingredient as to the follow-up procedures earlier in this chapter.

Many records must be updated resulting from approved change orders. Contracts between owner and contractor, subcontractor agreements and their respective job cost management files, estimating worksheets, cost budgets—all are examples of such records. Most important, the working plans and specifications must be updated to include details and notes that reflect the approved changes. This makes it easier for all field personnel to recognize the changes, thereby eliminating mistakes and confusion.

Problems

The general nature of change order work is a problem to a general contractor. Most of the common problems have been alluded to in this chapter. There are others: general contractor's errors and omissions in their estimates, overpricing from subcontractors, and owner-requested last-minute changes to changes, are examples of less common but nevertheless, occurring problems. A general contractor must deal with these events with professional acumen.

Chapter 17

Construction Commencement and Meetings During Construction

Commencement of Construction

Commencement of construction is an important phase of the project. The mode in which it is carried out makes a statement to the subcontractors, owners, design consultants, financing institutions, and other interested parties as to the character of the construction company and the manner in which it intends to build the project.

The commencement of a construction project should convey an enthusiastic and positive attitude by the entire general contractor's staff. Establishing these characteristics in the early stages of construction contributes to the development of good morale and a strong spirit as the job progresses. A cost-effective start-up should be implemented, even concerning the most menial tasks. This includes erecting project signs, setting up the field office, monitoring general condition items, and other related start-up work. If we are going to run a job that produces the desired results, let's begin in the beginning, and set the pattern for the forthcoming months or years of the project's duration. Commencement of the project should be planned with the same efficacy as is intended throughout construction.

Immediately after move-in, the field staff should scrutinize the working plans and specifications for errors, omissions, clarification items, and potential conflicts, and attempt to resolve these matters before construction moves into high gear. Also, a final review of the subcontract agreements should be accomplished at this time.

With the exception of supervision related to work that will be performed in the latter stages of construction, work such as finishes and punch list work, it is advantageous to have all supervisory personnel on the job when it starts. Employing this method affords them the opportunity to become thoroughly familiar with the plans, specifications, and progress schedules before the work they will be supervising commences. Additionally, they will be able to make their quantity take-offs, clarify and obtain answers to their questions, and otherwise prepare themselves for top performance. With these items out of the way, they will be free to spend more time in the field, applying their talents where they are most needed.

Meetings during Construction

Involved in every construction project is an extraordinary number of diverse events and operations. It then follows that an extraordinary number of individuals and firms are required to perform these multifarious tasks. As a result, vigorous differences of opinions concerning numerous aspects of the project are to be expected. It is imperative that job problems be resolved and differences settled to everyone's satisfaction, or, at least, to the overall benefit of the project. The principle expounded in the age-old adage, "a picture is worth a thousand words," but changed to "meetings during construction are worth a thousand letters and phone calls," when put into practice, produces the most effective means to this end.

As in preconstruction meetings (Chap. 14), agendas are prepared in advance, and the meetings should be held at times most convenient to all attending parties. Advance planning regarding each item to be discussed serves to capture and hold everyone's attention and generates the most productive meetings. These meetings serve many purposes—the three most important include the opportunity afforded all parties to freely express their concerns and frustrations; to resolve individual and collective problems and conflicts; and to obtain clarifications and directions. It is the general contractor's responsibility to conduct these meetings in a manner conducive to satisfactorily accomplishing these purposes.

General Contractor Staff

This personnel must not be burdened with unnecessary or nonproductive meetings which are time-consuming and detour energies that could be utilized in the field. Such waste is avoided by advance planning and preparation for each meeting. These meetings should be scheduled weekly, at the same day and time. To avoid job-caused interruptions and ensure that the attending personnel is fresh and not fatigued from the stress of the workday, the most beneficial time to hold the meeting is early morning, prior to starting the day's work.

Time allowed for the meeting should be of reasonable duration, and the subjects should be discussed in a crisp and to-the-point manner. In the event that excessive time is required to cover the agenda, the meeting should be continued at the earliest possible future date. Always avoid turning the meetings into a marathon, as boredom and inattentiveness will become evident. Notes covering the important items should be recorded and distributed to all personnel.

Job progress schedules

An in-depth review of job progress is hereby accomplished. The schedule is used to determine the status of all scopes of work and the previous week's work effect on the project completion date. Reasons for work items that are delinquent are discussed, and methods to make up time and improve their operations are explored and resolved. The necessity of updating the schedule is determined, and notes concerning weaknesses in its structure are recorded for reference on future projects.

Job cost

Printouts of actual job cost pertaining to labor, material, equipment, and all other items performed directly by the general contractor are distributed to the staff. After these costs are compared to the cost budgets, discussions ensue to ascertain reasons for any notable variances. In the case of cost overruns, methods and means to avoid their continuance are investigated and, when possible, corrected procedures are put into effect. Obviously, a happy atmosphere is evident with cost items that are under budget, and a minimum amount of time is devoted to this unusual, but satisfying, condition. In either event, the actual job cost is recorded in the space provided for in the budgets.

Subcontractor performance

Recognizing that subcontractors normally perform 85 to 90 percent of the work on a construction project, a large percentage of the time allotted for these meetings is on review and discussions concerning their performance. For this reason, the general contractor's staff must attend the meetings fully prepared to discuss specific items of concern relating to subcontractor performance. Notes or memos prepared in advance should be in order of importance and their contents explicit and precise.

Owner-Caused delays and problems

Late selections of colors and materials, untimely or unsatisfactory execution of work performed by firms employed directly by the owner, delays in forwarding pending change order information or executed change orders to the contractor, procrastination in processing and returning shop drawings, refusal to furnish required clarifications and details, uncooperative attitudes, improper or late payments of requisitions are several of the more typical owner-caused problems for which solutions are explored at the meetings.

Fundamentally, the owner is the primary party responsible for the aforesaid problems. However, it is a delusion to believe that this is always the case. Overzealous members of the contractor's staff, conscientious as they may be in the performance of their duties, frequently project combative demeanor and attitudes insensitive to the owner's concerns and dilemmas. These, and conditions having nothing to do with the contractor's staff, often influence unpropitious reactions from an owner. Meetings during construction are the opportune time to sort out the basic causes of these delays and problems and to search for corrective avenues to pursue.

Potential change order work

I repeat, a general contractor should not perform change order work prior to receiving formal, executed change orders, with the exception of work classified as emergencies. This statement does not mean that the contractor should not cooperate in every practical manner when aware of potential or pending changes to a project. When redirecting or delaying potentially conflicting work items will not result in job delays or entail extra cost, it is the contractor's obligation to act accordingly. Determining the likelihood of satisfactory achievement and the best method of procedure is accomplished at the meetings.

When it is reasonably certain that pending change order work will be approved, advance planning for the work is reviewed at the meeting. It is determined if required schedule adjustments, preliminary cost estimating, and other items that relate to the potential changes should commence immediately, or be delayed until a written change order request and related documents are received.

General job strengths and weaknesses

This portion of the meeting does not necessarily zero in on specific items, but is of a more general nature. These meetings exemplify, but are not necessarily limited to, the quality of performance by individual members of the field supervisory and main office management staff, personnel morale, and possible beneficial changes concerning equipment and methods employed.

All attending personnel are invited to express their feelings and ideas regarding any aspect of the project not specifically identified on the agenda. A profoundly important purpose, at this point of the meeting, is to placate emotions of hostility and frustration evident in individual staff members. Expressed understanding and working through the causes of these adverse emotions is best accomplished with the collective staff in attendance.

Contractor's and Subcontractor's Staff

These meetings are held with the general contractor's and all major subcontractor's staff in attendance, usually, on a weekly basis, at a designated time, in the general contractor's field office. Extenuating circumstances that require additional meetings with individual subcontractors are of common occurrence. The fundamental principles set forth earlier in this chapter under the heading of "General contractor's staff" are applicable to this meeting, and the subjects discussed are basically the same.

The primary functions the general contractor serves at these meetings are to ascertain and help work out the subcontractors' various job problems, furnish subcontractors with required clarifications, details, and answers to their questions, assist them in strengthening the weak areas of their work, and otherwise be supportive in every possible way. Enumerating the above items is simply for the purpose of establishing guidelines for these meetings as they have all been expounded in detail throughout this book.

Experienced general contractors are aware that subcontractors will, on occasion, attempt to negate, deny, and avoid their responsibilities. They may display a tendency to blame their problems on other subcontractors, owners, architects, engineers, the general contractor, or other parties involved in construction of a project. It is the general contractor's responsibility, agonizing as it often is, to determine when their complaints are justified or not, and to act accordingly. These meetings are the most fitting time to air and resolve such grievances.

Other important objectives of these meetings are to reaffirm the subcontractor's obligations to properly coordinate their work with other trades, display cooperative attitudes, and to maintain a standard of quality workmanship demanded by the contract documents. The general contractor should exert every effort to conduct these meetings in a manner that provides beneficial results, and replenishes the confidence and gainful spirit of all attending parties.

Contractor Staff, Architect, Engineers

These meetings are most effective when held with regularity. This avoids a buildup of unresolved problems and provides a steady flow of clarification, details, and information as needed throughout a construction project. They also serve to minimize misunderstandings and conflicts between the general contractors and the design consultants, and help to provide a mutual understanding of each firm's internal problems. In general, a better working relationship is realized, and a spirit of teamwork that benefits all parties becomes evident. Obviously, the frequency of these meetings is determined by the size and complexity of the project.

As in all other meetings during construction, before the meeting an agenda should be prepared enumerating the matters to be discussed in their order of importance. Matters pertaining to change orders, shop drawings, conflicts in the plans, consultant's and contractor's errors, quality control, schedules, and all other items of concern involving the consultants are targeted for discussion and resolution at these meetings. The responsibility to establish and maintain an amicable atmosphere and cooperative attitudes throughout discussions of each item lies primarily with the general contractor.

Contractor Staff and Owners

Normally, it is not as critical to schedule these meetings with the regularity of the above-mentioned meetings during construction. Subjects reviewed at these meetings are often not of a technical nature, but they require the efforts of all attending parties in order to resolve problematic issues, and to circumvent potential obstacles detrimental to the project. Astute, experienced contractors will take advantage of these meetings to maintain and improve owner-contractor relationships and goodwill.

Matters of importance most commonly discussed, and often debated, at these meetings pertain to progress delays, quality control, unresolved change order items, and various other complaints and concerns of both owner and contractor. Since the owner and contractor are equally dependent on each other to produce a satisfactory construction project, both parties must extend their concentrated efforts towards this goal, and avoid conflicts and combative attitudes throughout construction of the project. These meetings help to accomplish this harmonious goal.

Chapter

18

Quality Control and Punch List Work

Quality control and punch list work go hand in hand. The more effective that quality control is during construction, the less punch list work will be required at the end of the job. There are no fixed formulas from which cost estimates can be accurately prepared for these aspects of a construction project. A new set of problems and circumstances becomes evident with each new project. Experience, reference to cost control records from previous projects, and excellent quality control and punch list work supervision, are the three most effective tools a general contractor can depend on to provide accurate cost estimates and to keep expenses within the budget.

Quality Control

Controlling the quality of workmanship, materials, and equipment in order to realize conformance with the contract documents and the general contractor's standards, is the primary purpose of quality control inspections. These inspections must start the first day of construction and continue through substantial completion of a project. This applies to all major, and most minor, work items included in the structural, architectural, mechanical, plumbing, electrical, finish items, and site work phases of the job.

Correcting or redoing work in place is usually more costly and time consuming than performing the work originally. Supervision, labor, and equipment employed for this purpose is diverted from new work, and job progress is disrupted. Such problems are circumvented when the general contractor provides constant and astute surveillance to the work as it is being performed, and does not simply inspect the work after it is completed. Experienced general contractors have realistic goals: keep corrective work to a minimum—they realize that to avoid it completely is impossible—and have defective work corrected as soon as possible. Deferring corrective work, more often than not will entail greater expense and regrettable job conditions.

Enumerating the voluminous work items requiring quality control, common to most construction projects, is not the intent of this chapter, and would deviate from the purpose of the book. Actually, a comprehensive list identifying and commenting on each quality control item would constitute a book devoted solely to this purpose—one that I may write at a future date. However, the comments included in the following headings recapitulate the problem areas of quality control.

Subcontractor work

I realize the risk of redundancy is evident when repeating that subcontractors perform 85 to 90 percent of the work in a construction project. But I feel that this fact is so important as it reflects on subcontractor quality control that the risk is justified. Assigning inspections of subcontractor work to the assistant superintendents is the most practical and economical means to accomplish this task. But assistant superintendents have many additional responsibilities, and cannot always provide the time and effort these inspections require. When job conditions demand closer quality control supervision, I strongly advocate the employment of additional field supervisory staff for this purpose, even though incurring additional cost. This expense could be negligible compared to the potential damages suffered by the contractor due to inferior quality work.

General contractors must emphasize to their subcontractors, at the preconstruction meetings and throughout construction, that inferior quality will not be tolerated or accepted. When written and verbal requests to correct and improve inferior quality fails to produce satisfactory results, stronger measures must be taken. Reducing the noncomplying subcontractors' requisitions for payments, backcharging their contracts, and, in extreme cases, serving them with legal notices, are options general contractors must consider. It is often amazing how fast subcontractors will improve the quality of their work when faced with these conditions—that is, when they are responsible firms to begin with. If they are not, general contractors usually have only themselves to blame.

Contractor-performed work

Methods and standards with respect to quality control are elucidated at the general contractor staff preconstruction meetings (see Chapter

14). Throughout construction, they are reviewed and reinforced during job site inspections and at field meetings. Good contractors will extend the same degree of effort to produce quality work, as it relates to their directly performed work items, as they do to controlling expenses for these same items. Controlling quality and controlling expenses are of equivalent significance.

The same diligent inspections and supervision apply to work performed directly by the general contractor as to that performed by the subcontractors. In most cases, contractors can control the quality of their work with less effort than necessary to enforce quality control for work performed by subcontractors. General contractors must set an example, thereby not affording subcontractors an opportunity justifiably to compare their inferior quality to that of the contractor-performed work. In other words, the maxim "do as we say, not as we do" does not produce satisfactory results on a construction project.

Cost estimates

The word *perfection* is not included in a general contractor's vocabulary. In spite of extensive efforts to produce a good quality job, additional funds will be required for corrective work. This means that an appropriate amount must be included in the job cost estimate for this purpose. To estimate this amount accurately is difficult, but experienced estimators, particularly those with field experience, are able to approximate these costs with surprising accuracy.

On what should be rare occasions, general contractors are burdened with the additional cost of correcting work performed by subcontractors. This condition may arise due to ambiguity in the contract documents, misdirection by the general contractor, or damage to work by unidentifiable parties. In these instances, and when the expenses are minor, the contractor may find it more prudent to pay for the corrections than spend the time necessary to collect the charges from the subcontractors, owners, or design consultants. The estimating staff should include a small percentage factor in the job cost estimate to cover such expenses.

Unreasonable owner requests

On occasion, owners and their agents may make unreasonable demands with respect to quality control. They may choose to ignore allowable tolerances, or expect more than that to which they are entitled. When this happens, general contractors should exert every effort to find solutions satisfactory to all parties. Failing this, the contractor's best defense is reference to the contract documents, and when necessary, acceptable standards of the industry. Obviously, before taking this position, the contractor must be totally confident that the work in question was properly performed, and satisfies all of the specified quality control requirements.

Punch List Work

In this book, punch list work is defined as the final repairs, adjustments, replacements, or any type of corrective work required to satisfactorily complete a construction project. Not to be confused with work performed subsequent to completion of a project, under the auspices of warranties and guarantees, punch list work commences as soon as each portion of each scope of work reaches an early stage of completion.

Effective execution of punch list work is a long and arduous task that necessitates the combined efforts and cooperation of the general contractor, subcontractors, and owner's agents. The pressures to complete a project on schedule and within a budget correlate with the quantity of punch list work required. Hence, the importance of establishing and maintaining excellent quality control while the work is being performed is emphasized. Better quality control means less punch list work.

Supervision

Experience, persistency, tenacity, patience, detail-oriented—these are the characteristics indispensable to a proficient punch list superintendent. Making inspections, recording items needing work, distributing work lists to subcontractors, monitoring the work, and follow-up are all included in the responsibilities of these people. At times, it can seem like a never-ending job, undeniably repetitious, frustrating, and boring. Nevertheless, it is a job requiring, in addition to the aforesaid qualities, substantial knowledge of all phases of construction, perceptive judgment, and judicious execution.

Of equivalent importance to the already mentioned responsibilities of a punch list superintendent is the ability to satisfy the owner. The superintendent must be capable of determining when the owner's requests for corrective work are justified, or when demands are excessive and unreasonable. Knowing when to cross the fine line and perform additional work to satisfy an owner demonstrates the judgment inherent in a good punch list superintendent. Finally, the superintendent is depended on to act as the mediator for those conflicts that arise between subcontractors concerning responsibility of punch list work.

When to start

I want to take this opportunity to clearly distinguish quality control from punch list work, and will use the example of a reinforced, poured-in-place concrete column to serve this purpose. During erection of the forms, inspections are conducted to ensure that the forms are of correct dimensions, tight and properly braced; that acceptable material is used; and that all other aspects of the form work will produce a plumb and structurally sound column. The concrete is inspected for proper proportions and types of ingredients, conformance to the design strength, and the time lapse from dispatching to placement. The reinforcing steel is inspected for conformance to the approved shop and placing drawings and for adequate clearance between the steel and the forms. All of these inspections take place prior to placing the concrete. During concrete placement, the methods and equipment used for vibrating are checked to assure that adequate, but not excessive, vibrating is employed. All of these inspections are defined as quality control. When written out, they may seem excessive. However, in the work day of a structural superintendent, they are considered mundane duties.

Now to explore the punch list work related to this same column. Subsequent to stripping the forms, inspections are conducted concerning honeycombing, exposed reinforcing steel, bulges, and dimensions—all of the items that effective quality control should avoid. When these, or other, defects are evident, corrective actions including the necessary rubbing, grinding, chipping, patching, or even demolishing and replacing are promptly put into action.

The question of when to start punch list work is answered: as soon as possible. The example of the concrete column is simple in that once the forms are stripped, the work prior to punch list items is completed. Other more complex scopes of work that have multiple component parts require punch list work as independent phases are completed, as in the example of dry wall work. This scope of work, performed by one or more firms, includes studs, insulation, blocking and backing, drywall board, taping, and finishing.

Each one of these work items must be completed, including punch list work, before the next phase can be completed. In this example, there are five work items which must be inspected, and required punch list work completed, before finishing the drywall scope of work. The sequence of events eliminates the luxury of electing when to commence the punch list work. It must start and finish immediately upon completion of each component work item. Allowing as much time as possible to perform punch list work is a prudent decision, and one practiced by experienced general contractors.

There may be contractors who will argue, and with some justification, that these two examples are classified as corrective work, not punch list work. Their examples of punch list work might be painting touch-ups, door lock adjustments, or similar work items. I must agree. After all, I did define punch list work as final corrective work. However, my objective is to impress upon the reader that punch list, or corrective work—whichever term you prefer—should start as soon as possible.

Subcontractors

For the most part, subcontractors deplore punch list work. Consequently, their attitude is frequently remiss as to starting and satisfactorily finishing it. Oftentimes contributing to this factor is their failure to include in their cost estimates sufficient labor and material for this critical portion of their work. Even experienced subcontractors will erroneously gamble that their required punch list work will be minimal. This does not diminish the need for satisfactory performance, nor does it excuse or reduce the subcontractor's responsibilities. It does, however, make the general contractor's work more difficult. These facts emphasize the need to achieve the best possible quality control, thereby reducing punch list work.

Enforced early starts, efficient inspection and supervision, and an unyielding determination to achieve timely and satisfactory performance, all administered by the general contractor, are the three most effective ingredients essential to realizing acceptable punch list work from subcontractors. Supporting these ingredients with early distribution of properly prepared work lists and constant follow-up will serve to reduce the agony common to subcontractor punch list work.

As the end of a project approaches, increased pressure to complete minor punch list work items becomes more apparent. A few examples include adjustments to finished hardware and millwork items, minor caulking and sealant work, realigning electrical switch and receptacle plates, installing missing screws and anchors for medicine cabinets and bath accessories, straightening loose support grids for luminous and acoustical tile ceilings, and of course, final cleanup of trash. These are just a few examples. There are many more.

At this point of the job, the subcontractors are usually struggling to complete their major punch list items, and their efforts to complete the minor work is neglected. Since these are simple work items easily performed by the general contractor's staff, often during the course of their final inspections, the contractor may exercise good judgment in deciding to complete this work with company personnel. This is easily accomplished by simply filling a common shopping cart with screws, bolts, sealants, and other miscellaneous items that will accompany the inspecting personnel and their assistants. For example, when a switch plate is out of square, this personnel can straighten it in a fraction of the time required to note it on a list, send the list to the field office for distribution to the electrician, and follow up with still another inspection to ensure that the switch plate was straightened. When the contractor's expenses for these procedures are reasonable, and are properly documented, they can usually be deducted, with the subcontractor's approval, from the final sums due to the responsible subcontractors.

Work lists

These lists are prepared simultaneously with the punch list inspections. Properly structured lists are comprehensive in their contents with precise descriptions and locations of the work required. For example: Install return air grill, apartment 11C, master bedroom, north wall. Copies of the master list identifying all necessary punch list work, or excerpts from these lists, pertaining to individual items are transmitted to the subcontractors and other responsible parties.

Forms for these lists are most effective when modified to adapt to the peculiarities of each project. Dates that inspections are made, dates of transmittal, and signature of the inspecting party are recorded on each list. The lists are updated with each repeat inspection and forwarded to the responsible parties. During each inspection, work that has been completed is checked off, and necessary remarks are entered in space provided on the forms. An excellent software computer program designed to produce these, and all other contractor required forms, is called *Formtool*®, published by Bloc Development Corporation of Miami, Florida. These work lists are an essential tool when conducting inspections, advising subcontractors of their required punch list work, follow-up, and support information, and—if necessary—supporting legal action. They should be treated as equally significant to all other written documents utilized on a construction project.

Follow-up

Completion of punch list work requires persistent follow-up by the general contractor. This requirement applies to carrying out inspections, updating and revising work lists, and continuing the efforts to have the subcontractors actually complete their punch list work. There is never too much follow-up with respect to punch list work—usually, not enough.

Punch list work is one of the agonies inherent in the life of a general contractor. In a manner similar to a prize fighter who shows signs of fatigue in the late rounds of a fight, the general contractor's and subcontractor's personnel grow weary of the repetition and length of time characteristic of punch list work. It starts early in the job and picks up intensity as the project nears completion. Thorough job preparation, superior quality control, use of experienced subcontractors, and proficient field supervision—all furnish an easier way to accomplish this aspect of a construction project.

Chapter

19

Final Documents, Payments, and Performance After Completion

Final Documents

Prior to receiving final payments, the general contractor must submit several documents, drawings, and data to the owner. This obligation is no less binding on the contractor than is completing the project, and must not be performed in a casual manner. As the project approaches completion, this material should be assembled and checked for accuracy and conformance to the contract requirements, prior to transmitting it to the owner. Transmittal sheets that identify each item, are dated and attached to the documents, with space provided for the owner's signature and date of receipt. Copies of these documents are kept in the contractor's permanent files.

As-Built drawings

These drawings indicate changes from the original working drawings incorporated in the project throughout construction. They serve the important purpose of helping accurately to locate problems that may arise, and are referred to when performing future work. They are especially critical when dealing with structural items and mechanical piping located underground or within walls, ceilings, and floors.

Warranties and Guaranties

The project specifications enumerate those items requiring the furnishing of warranties and guaranties. Letters, certificates, catalogs,

113

and brochures typify these documents. They are written pledges to repair or replace material, equipment, or workmanship that becomes defective within the warranty period. They also include instruction manuals, maintenance information, and delineation of design characteristics of equipment and specialized materials.

A common problem concerning these documents arises when determining if the warranty and guaranty periods pertaining to certain items, begin when specific work items are completed, or when the entire project is completed. For example, when the entire structural scope of work is completed, inspected, and approved months before the balance of the job is completed. Do the warranties and guaranties concerning this work commence immediately after its completion, or not until completion and acceptance of the entire project has been accomplished? From the owner's point of view, warranty periods with respect to all scopes of work will commence at the latter date.

The area of contention becomes evident when the subcontractor responsible for the structural work refuses to perform corrective work, on the grounds that the warranty period, as measured by the completion date of this scope of work, has expired. An interpretation satisfactory to all interested parties cannot always be realized when it depends on the project specifications, general and supplementary conditions, and other related documents. The most judicious way to avoid this problem is for the general contractor to clearly state, in all subcontract agreements, that the effective beginning dates concerning warranties and guaranties coincides with completion and acceptance of the entire project.

Certificates of occupancy

Temporary certificates of occupancy for completed areas of a building may be issued by the building department. This allows occupancy restricted to the approved areas only and usually requires temporary barricades and other protective measures. The certificates are valid for a limited time, and become null and void when the balance of construction is not completed, inspected, and approved by the date set forth on the certificates. These certificates are delivered to the owner upon issuance to the general contractor.

Final certificates of occupancy are issued when construction is completed, subsequent to final inspections and approvals by all governmental agencies. These inspections include building, zoning, mechanical, plumbing, electrical, fire, and others that may be required by the local building department. These final certificates of occupancy are included with the documents to be delivered to the owner at completion of a project.

Payments

I have emphasized, throughout this book, the importance of perpetuating a spirit of cooperation, goodwill, and trust between the owner and the general contractor. At no point of the project is this principle more beneficial to the general contractor. Owners often question the contractor's intent to perform corrective work that may arise after they have received final payment. Contractors, while being sensitive to these concerns, must remind the owner of their favorable performance throughout construction, and reassure the owner that this performance will continue subsequent to completion and receipt of final payments. When sincerity and truthfulness is evident, it often becomes an easier task to collect final payments.

During the course of a general contractor's career, there more than likely will be situations whereby the owner will withhold final payments for reasons that reflect bad intent, having nothing to do with the aforesaid owner's concerns. Retaliation to satisfy harbored hostile feelings, which result from conflicts during construction not resolved to the owner's satisfaction, or unjustifiable attempts to negotiate the final amounts due the contractor, for the purpose of saving money, are classic examples of these situations. In the face of these occurrences, when contractors, with no success, exhaust every reasonable means to collect final payments, they are left with no alternative but to initiate legal action and refuse to perform future work.

Final checking

Before requesting final payments, it is the general contractor's obligation to thoroughly check the entire project to ensure that all work has been properly completed. Do not expect the owner to make final payments even when the insignificance of the incomplete work is apparent.

Support data

A general contractor's requisition for final payment will require accompaniment of supplementary support data, including the various affidavits, release of liens, and other specified documents. This data is of a legal nature and, therefore, must be properly prepared and executed on the specified forms. Failure to adhere to these requirements is valid reason for the owner not to make final payments.

Owner payments to subcontractors and suppliers

In certain circumstances, the owner or general contractor may elect to have the owner make final payments due to subcontractors or suppliers and to deduct these amounts from the contractor's final requisition. When this occurs, both the owner and contractor must be in agreement concerning the methods that will govern the payments. Both parties must be assured they will receive the same degree of protection as if the contractor were making the payments.

This method of making payments to subcontractors and suppliers serves three important purposes. First, it furnishes the owner with the security that subcontractor and supplier claims and liens, resulting from the contractor's failure to make proper final payments, will be avoided. This is of particular value when general contractor performance and payment bonds were not required. Secondly, in the case of strained relationships between the general contractor and subcontractors, the subcontractors will sometimes complete their work more expeditiously, when assured that the owner will pay them directly. Finally, this method of making final payments relieves the contractor of having to furnish the funds in advance of being paid by the owner.

There are disadvantages to these procedures. The most serious is the general contractor's potential loss of control of the subcontractors. When owners make payments directly to subcontractors, the owners sometimes become the commanding force, in lieu of the contractor. Also, the owner, as a result of assuming this responsibility, may lose an appreciable degree of respect for the general contractor. A contractor should carefully consider these facts before agreeing to this method of making final payments to subcontractors and suppliers.

Performance after Completion

Completion of a construction project, acceptance by the owner, and receipt of final payments, do not constitute conclusion of the contractor's obligations. The contractor must perform all required work under the auspices of warranties and guaranties in the same enthusiastic manner displayed throughout construction. Before assuming that this will be no problem, be aware that the warranties and guaranties are in effect for long periods of time. Experienced general contractors are prepared to honor these obligations, and include contingency sums in their cost estimates to cover their related expenses.

The practice of performing additional corrective work beyond the requirements of warranty and guaranty periods is a sure way to enhance a general contractor's favorable reputation. It will serve to promote recommendations from owners, design consultants, and financial institutions. This is not meant to imply that contractors should supply free maintenance to a project, or allow advantage to be taken of their good intent. It does mean that contractors should examine the benefits involved, and make prudent business decisions in this regard.

Chapter
20
Potential for Growth

General contractors who have successfully employed the principles set forth in this book, and whose goal is to develop greater volume and economic growth, have four fundamental avenues of growth potential to explore in their pursuit of this goal: new types of projects, offices in new locales, construction management, and equity positions.

There are significant risks that accompany all four areas of growth potential. The contractor must not only be aware of these risks, but have the foresight and business acumen to carefully analyze and measure the advantages and disadvantages, before leading the company into these new horizons of expansion. Expenses and management staff efforts increase immediately after implementation of the procedures necessary to achieve greater volume and earnings. This increase becomes particularly apparent concerning new locales. Once these procedures are out of the starting gate, the expanded cost and efforts cannot be recouped, and reversal of activities will result in losses, sometimes substantial ones. Obviously, a contractor's risk is of a measure proportionate to the company's growth ambitions. However, the nature of the business requires larger percentages of capital expenditures than that of many other businesses.

Types of Projects

Many general contractors, when seeking new work, will concentrate their efforts in obtaining contracts for the type of projects most familiar to their company. This is most evident when contractors specialize in certain type projects. It is a prudent and effective policy, but it does limit growth potential. Qualified contractors who desire to strive for greater company growth can pursue the various types of projects expounded in Chap. 2 ("Types of Projects"), particularly in metropolitan areas.

A general contractor should not seek contracts for new types of projects until completely satisfied that the company staff is fully qualified to efficiently execute all aspects of the project, especially the estimating, management, and field supervision departments. When these qualifications are not apparent, contractors should postpone their entry into construction contracts for types of projects unfamiliar to the company. Also, it is unwise to burden the existing staff with the additional responsibilities of a new type of project, when it will decrease their efficiency with respect to jobs under contract, or to familiar type work available for bidding or negotiating. This theory will allow a general contractor to pursue a new type of project in a conservative manner and with minimum start-up expenditures.

For the contractor with substantial financial reserves, I recommend a more aggressive approach, one which entails developing a separate management, estimating, and field supervisory staff for the primary purpose of building new types of projects. This provides the means for a general contractor to approach a new type of work with the same degree of confidence and stability as when seeking and building familiar projects. It also reduces the risk of overburdening existing personnel, and most importantly, serves to gain the confidence of owners and design consultants.

New Locales

Opening offices in new locales presents the greatest risk, requires the most effort, and necessitates the largest capital investment of the four avenues of growth potential alluded to in this chapter. It also offers the greatest opportunity for growth in the least amount of time. Before pursuing this complicated major avenue of growth, a general contractor must be absolutely sure that it is in harmony with the philosophy and long term goals of the company, and that it engages the necessary energy levels and ambitions of the company executives and management personnel. Unless a positive response to these basic and significant items is apparent, expanding operations to new locales is not justified, despite all other conditions and prospects being favorable.

The decision to open offices in new locales must be preceded by a thorough investigation of the real estate activity, overall economic stability, and long-term growth potential as it relates to the construction industry. These investigations should be performed by professional firms that specialize in such research and investigation. Other items of major significance to be explored are the availability of qualified management personnel at all levels, existing labor force, and experienced subcontractors and suppliers. Extensive investigation that includes interviews, checking references, speaking to owners, consultants, and lending institutions, and a general comprehensive research program concerning these items should be conducted by the general contractor's executive staff. Additionally, the company attorney should review the laws pertaining to general contractors to ascertain any substantial differences between the existing and the proposed new locales.

The temptation to open a permanent office in an area temporarily booming with construction activity must be resisted when research and investigation clearly indicate that the current activity will be followed by severe and extended slowdowns. A general contractor can take advantage of the temporary activity by applying the unwritten laws of conservatism when moving into this type of area. In other words, avoid a full-blown organization and operation that will be costly to disband after a reasonably short time. Renting instead of buying equipment, working out of modest executive offices, using the facilities and staff from the main office whenever possible, and preparing only for the present, exemplify judicious judgment when opening temporary offices in a temporary locale.

Conversely, when investigations indicate favorable long-term conditions, the new locale should be penetrated with ambition, vigor, and optimism. The necessary capital investments must be furnished, and a full compliment of personnel made available to set the tone for a long and prosperous venture. Be aware that general contractors who have the sagacity and expertise to run a successful construction company in one area, are qualified to repeat this performance in another. Make a statement: "This company is here to stay, and to become an important part of the community."

When a general contractor moves into a new area, it is similar to an employee being transferred to a new city to open a branch office. Adjusting to new surroundings, making new contacts, becoming familiar with the firms and individuals you will be doing business with are all essential elements. Introducing their company and establishing favorable relationships with the local architects and engineers should be accomplished by those members of the contractor's management staff who are experienced in working with design consultants and who display a professional demeanor. Company brochures, informative news releases, and letters of recommendation from clients and consultants serve as valuable support material when asserting the company's qualifications during these introductory meetings.

The same principles apply to subcontractors and suppliers as to architects and engineers, with two basic differences. First, the general contractors, project managers, and superintendents should be involved in the introductory meetings, and secondly, review of the subcontractors' and suppliers' qualifications are the dominant points of significance. Subsequent to meeting these new firms, the general contractor's staff, through further investigation, must become assured that they complement the established standards of the general contractor before including them in their list of qualified subcontractors and suppliers.

It is essential that a new office be properly staffed and have all systems in operation so as to enable its daily activities to be independent of the main office. However, ongoing communication must be maintained with the main office to ensure minimum deviation from the company's basic policies and mode of operations. Also, frequent communications provide a way to gain beneficial knowledge from the current experiences of each office. Personnel employed for the new office benefit during their breaking-in period when they are free to communicate with the main office for assistance and support. Computers, fax machines, and high tech telephone systems provide easy, fast methods of communication between offices, and are necessary tools when operating in multiple locales.

Opening an office in a new area requires high risk and a large capital investment. Therefore, it is unthinkable to have anyone but the most experienced, qualified, and dedicated person to be in charge of, and responsible for, all aspects of a company's operations pertaining to a new office. This necessitates compensation commensurate to the work and responsibilities involved. Realizing that the success of a new office is greatly dependent on this person, the general contractor should consider offering more than just a substantial salary. Profit sharing, stock options, or other means of financial rewards will help to enforce the evidence of loyalty and hard work essential to successfully managing a general contractor's business.

Construction Management

Here, I tread on delicate ground. But, if I am to be honest with my readers, I must write it as I believe it—and I do not believe construction management firms are as valuable as they are popular in the construction industry. Construction management is a relatively new term used in the construction industry to identify firms which perform age-old functions. Construction managers often begin their work in the preconstruction stages, starting with the selection of the design consultants, and working with the consultants throughout development of the plans and specifications. They prepare cost budgets, preliminary network construction schedules, and bidding documents. They also act as the qualifying agents regarding general contractors and subcontractors, negotiate construction contracts, specify forms and documents to be used throughout construction, suggest value engineering changes, and undertake a multitude of other preconstruction items.

Once construction has commenced, the construction manager continues to act as the owner's representative with respect to all phases of construction. The construction manager inspects all work performed, reviews payment requisitions and change orders, processes shop drawings, acquires needed details and information from the design consultants, and oversees the many other diverse items inherent to a construction project.

When analyzing these multifarious functions of a construction manager, the reader might come to the conclusion that my belief concerning their importance is erroneous. Emphatically, I proclaim that expert owner representation regarding all of the aforesaid items is important, but I question the wisdom of paying the cost and fees charged by construction management firms when more economical and efficient means can be employed. In addition to their fees, these firms are reimbursed for their out-of-pocket costs. This includes key personnel and support staff, their travel and field office expenses, all of which amount to substantial sums.

Performance of a construction management firm's job can be accomplished with equal, and frequently more effective, results by personnel working directly and solely for an owner. Often, this method requires a smaller staff to accomplish the same work, thereby reducing costs and eliminating fees. Also, a greater degree of loyalty may be evident than in contracting with a construction management firm, which may have multiple projects under contract. If a general contractor is not used, and all work is subcontracted, the construction management firm assumes additional responsibilities, and their fees are based accordingly. A fully qualified person with experience in managing construction projects is capable of hiring the necessary support staff to satisfactorily perform the tasks of a construction management firm when this condition prevails.

One might ask, doesn't the same principle apply to general contractors? The answer is, not nearly to the same extent. General contractors are legally bound to complete a project at a cost not to exceed the contract amount and, even in the case of a cost plus a fee contract, they must build the project, not just manage it. Construction managers must use their best efforts to perform under the terms of their contract, but usually are not encumbered with the obligations and responsibilities of a general contractor.

I have witnessed situations where owners have contracted with construction management firms for no other reason than that they felt it was the thing to do. I refer to projects where working drawings and specifications were provided, and the general contractor was conscientious, experienced, and furnished a 100 percent performance and payment bond. Questionable items, change order work, requisition review, and other conditions common to a construction project could easily have been handled and processed with the architect acting as the owner's agent, in lieu of a construction management firm. I have also been witness to the unfortunate circumstance, whereby construction management firms, sometimes in their efforts to impress an owner, and other times due to lack of qualified personnel, have complicated simple matters and caused strained relationships between owners and general contractors.

Now that I have done a thorough job of alienating construction management firms, and for that matter, people who work for my company in that capacity, I will attempt to outline how construction management can be a productive avenue of growth potential for a general contracting company. Before proceeding, I will take this opportunity to advise the reader, that while we are primarily general contractors, throughout the years we acted as construction managers on many large projects—profitably, I may add. Given the opportunity, we will continue to do so. Therefore, I hope my preceding comments that clearly question the need of a construction management firm on a project are not construed to be self-serving.

A general contractor should be aware that all of the responsibilities delegated to a construction manager prior, and subsequent, to construction, can be effectively fulfilled by a qualified general contractor. A construction manager cannot always perform all of the general contractor's responsibilities, particularly with respect to actual construction aspects, with the same degree of efficiency. Simply translated, this means a contractor who elects to pursue growth potential through construction management, does not have to learn new systems or techniques, or furnish additional training to company personnel. Certain methods and philosophies will require modification, but this is of minimum consequence.

A contractor will have to choose between being a general contractor or a construction manager on any given project, as a construction firm cannot perform in both capacities on the same project. Different fee structures come into effect, obvious conflicts of interest become apparent, and professional ethics would be violated. And of course, this would defeat the purpose of having an owner's representative body acting as the agent between the owner and contractor. This does not mean to imply the same personnel cannot be used in either capacity, however, different types of records and procedures will be required.

Finally, the advantages and disadvantages of construction management compared to general contracting can be stated with brevity—less capital investment, less risk, and less profit.

Equity Positions

To exemplify the significant value of equity positions with respect to increased volume, I will relate the events our company experienced on a large project we were negotiating in competition with two other general contractors. Negotiations pertained to the cost and the fee on an upset price contract. After lengthy negotiations, it became apparent that our proposed price and fee were very close to that of our competitors'. Since all three contractors were prequalified before being invited to submit their proposals, all three had extensive experience in this type of work, and enjoyed equally good reputations, it almost became a flip of the coin for the owner.

It was at this point of negotiations that we suggested to the owner that we forego our fee, and instead take an equity position for its equivalent value. The sum that would be waived regarding the fee was not of primary importance to the owner. What did sell him, was his feeling that, as owners, we would do all in our power to build a project in the most cost-effective, expeditious, and proficient manner. We would have done that in any event, but our suggestion provided the owner with the additional security and comfort he needed to award us the contract. We employed this method on future projects with a similar degree of success.

As to a general contractor's financial growth potential related to equity positions, there are several factors to consider. A contractor's fee or anticipated net profit is, usually, surprisingly small in comparison to the construction cost of a project. Equity positions on successful projects can produce far greater earnings over the long term than the contractor can expect from normal fees and profits. Another important consideration is the tax consequence a contractor may face on the profits from a construction contract, versus return on investment related to an equity position. Even under the current tax laws, I believe the latter is more beneficial taxwise. Of course, this would have to be confirmed with the contractor's accountant on a project-by-project basis. There is the additional advantage of gaining the owners' confidence, not only as their contractor but as their associate. This provides a general contractor with a favorable edge when being considered by owners for new projects.

General contractors should not enter into an equity position arrangement without first investigating the economic value, the owners' intent, and the potential risk of the venture. While there are no guarantees that any business venture will be successful, general contractors should not rely solely on an owner's judgment, but should use their own business acumen in determining the economic value of a project. Owners may go forth with a project knowing that return on investment and long term appreciation will be negligible, but their intent to proceed is based on their tax position. This may make it a viable project for them, but not necessarily for the contractor. The general contractor must carefully analyze the potential risk as it relates both to financial liability and to possible adverse effects on a contractor's reputation. The risk factors may not affect an owner with the same degree of severity as the contractor.

Serious unfavorable consequences can become evident when a general contractor takes an equity position with the attitude that there is nothing to lose but the profit that would have been generated from the construction contract. If contractors are not able to ascertain the potential upsides and downsides of a project with the same degree of certainty as though they were the developers, they should not consider equity positions. I strongly advocate equity positions for the long-term economic growth of general contractors, but only if they have the financial resources to defer income gained from fees or profits, the ability to research and investigate the aforesaid aspects of a project, and the assurance that owners, with whom they intend to become associated, are reputed to display the utmost veracity in their business dealings.

Finding a qualified owner with a project that meets with the contractor's approval, and who is receptive to equity positions, is difficult. Not all owners need the reassurance that better performance will be realized if the contractor has a financial investment in a project. And it is a rare occasion that an owner needs the support of a contractor's fee to proceed. Also, in many instances, lending institutions require general contractors to provide performance and payment bonds, and to be a separate entity from the owner. Nevertheless, on occasion, equity position arrangements are available. When the opportunity presents itself, the general contractor should be receptive to exploring this avenue for growth potential.

Being awarded a new contract, realizing projected profits on a project, entering new avenues of growth potential, watching the company reach its goals, and being proud of your profession are the rewards of being a general contractor.

Appendices

Appendix A	Subcontractor Requisition Form	127
Appendix B	Partial Release of Lien (Subcontractors and Suppliers)	129
Appendix C	Final Waiver of Lien (Subcontractors and Suppliers)	131
Appendix D	Subcontract Agreement	133
Appendix E	Requisition Printout (General Contractor to Owner)	153
Appendix F	Subcontractor Status Report Printout	161
Appendix G	Job Cost Management Printout: Budgets and Actual Costs (Labor, Material, Rental Equipment, Miscellaneous)	165
Appendix H	Job Cost Management Printout: Volume and Dollar Amounts (Labor, Material, Rental Equipment, Miscellaneous)	211

Appendix A
Subcontractor Requisition Form

(FILL IN SUBCONTRACTORS NAME & ADDRESS)

REQUEST FOR PAYMENT

REQUISITION No. (   ) DATE ....

Millman
CONSTRUCTION CO. INC.
GENERAL CONTRACTORS
1301 Dade Boulevard, Miami Beach, Florida 33139
Telephone (305) 556-6582

PROJECT:

OWNER:

GENTLEMEN:
THIS REQUEST FOR PAYMENT IS FOR WORK PERFORMED ON THE ABOVE PROJECT THRU THE PAY PERIOD ENDING (   )

DO NOT WRITE IN THIS SPACE

ORIGINAL CONTRACT AMOUNT	$
APPROVED CHANGE ORDERS	$
EXTRAS	$
CREDITS	$
TOTAL REVISED CONTRACT	$
VALUE OF WORK PERFORMED TO DATE (per breakdown attached)	$
VALUE OF MATERIAL STORED ON SITE	$
TOTAL.	$
LESS _____% RETAINAGE	$
AMOUNT EARNED TO DATE	$
LESS PREVIOUS REQUISITIONS	$
AMOUNT OF THIS REQUISITION No. ___	$

RELEASE

The Subcontractor certifies that all materials, labor and services furnished by him through the above mentioned pay period have been fully paid for (Except as listed below) and the premises of the above named job cannot be made subject to any valid lien or claim by anyone who furnishes material, labor or services to the Subcontractor for use in said job; and the Subcontractor hereby Releases MILLMAN CONSTRUCTION CO., INC. General Contractors, and the Owner, from any further liability in connection with all materials, labor, and services furnished by the Subcontractor thru the pay period.

This Release is given in order to induce payment in the amount of $_____ and on receipt of said payment by the Subcontractor this Release becomes in full force and effect.
EXCEPTIONS ARE AS FOLLOWS:

STATE OF:
COUNTY OF:
SWORN TO AND SUBSCRIBED BEFORE ME THIS
_____ DAY OF _____, 19 ____

SEAL
NOTARY PUBLIC

....
SUBCONTRACTOR
BY: ..../......................
TITLE: ....
DATE: ....

SEAL

Appendix B
Partial Release of Lien
(Subcontractors and Suppliers)

PARTIAL RELEASE OF LIEN

THIS IS A PARTIAL RELEASE ONLY AND CANNOT BE USED AS A FINAL RELEASE

The undersigned_____, for and in consideration of the sum of Ten Dollars ($10.00) and other good and valuable considerations to the undersigned in hand paid, the receipt whereof is hereby acknowledged, does hereby waive, release, remise and relinquish the undersigned's right to claim, demand, or impose a lien or liens in the sum of $_____, for work done or materials furnished or any other kind or class of lien whatsoever on the following described real property in Dade County, Florida.

General Contractor: Millman Construction Co., Inc.
This agreement constitutes a Partial Release of Lien by the undersigned in sum of $_____, and shall not operate to release any lien of the undersigned in sums in excess of this Partial Release or for labor, services and materials to be furnished after this date, and is given in accordance with the Florida Mechanics' Lien Law.

Witness the hand and seal of the undersigned this _____day of _____,A.D., 19____.

Signed, and sealed in the presence of:

PLEASE DATE ALL RELEASE OF LIENS

_____

_____
(Contractor, materialman, subcontractor, etc.)

_____
Title:_____

KINDLY FOLLOW INSTRUCTIONS: Partial Release of Liens, must be completed in full detail, signed by authorized person, corporate seal affixed, signed in the presence of Two (2) witnesses.

RETURN IN DUPLICATE AT ONCE TO:

MILLMAN CONSTRUCTION CO., INC. 1301 Dade Boulvard, Miami Beach, FL 33139

Appendix C
Final Waiver of Lien
(Subcontractors and Suppliers)

Appendix C

FINAL
WAIVER OF LIEN

KNOW ALL MEN BY THESE PRESENTS, that _____, for and in consideration of $_____ and other good and valuable considerations, lawful money of the United States of America to (him) (it) in hand paid, the receipt whereof is hereby acknowledged, does hereby waive, release, remise and relinquish any and all right to claim any lien or liens for work done or material furnished, or any kind or class of lien whatsoever, on the following described property:

That the undersigned, _____, individually and/or as _____ (officer) of _____, a corporation, in order to induce MILLMAN CONSTRUCTION COMPANY to make payment in the sum above set forth, the receipt of which is hereby acknowledged, does affirm and warrant that all sub-contractors and materialmen and/or laborers of _____ have been paid for services, labor and materials delivered to or performed at the job site above referred to and that there are no such persons, firms or corporations through your Affiant who could claim a lien on the premises by reason of materials delivered, services performed or labor performed. The undersigned affirms that the value of services performed or materials delivered to the job site is equal to or exceeds the sum herein receipted for.

  Your Affiant further states that there is below listed names of all persons, firms or corporations who have heretofore performed and/or delivered materials for _____ _____ at the job site:

  IN WITNESS WHEREOF, (I) (We) this day set our hands and seals at Miami, Dade County, Florida, this ____ day of _____, 19

Signed, Sealed and Delivered in the
Presence of:

                  _____.

_____   _____

                  _____

_____   _____

                  _____

STATE OF FLORIDA)
COUNTY OF DADE)

  BEFORE ME THIS DAY personally appeared _____, _____, and _____, persons to me well known who, if signing in the capacity as an officer of a corporation, acknowledged that he had the corporate authority to do so and that he did so for the use and purposes therein set forth and/or who acknowledged that if he signed in the capacity as an individual owner, that he did so for the uses and purposes therein set forth, and in any event, who stated that the matters and facts herein set forth are true and correct.

    DATED _____.

                 _____
                 Notary Public, State of Florida at Large

                 My Commission Expires:_____

Appendix D
Subcontract Agreement

MILLMAN CONSTRUCTION CO., INC.
1301 Dade Boulevard
Miami Beach, Florida 33139
Telephone: 531-0513

SUBCONTRACT AGREEMENT

THIS AGREEMENT,

Executed this   day of   , 1985, by and between:

Millman Construction Co., Inc.,
a Florida corporation,

hereinafter called the "Contractor", and

hereinafter called the "Subcontractor"; for the following described project:

4000 Island Boulevard
Williams Island, a Condominium
located on Island #2, Williams Island,

hereinafter called the "Project"; being constructed for:

4000 Island Boulevard Association, Ltd.,
a Florida limited partnership,

hereinafter called the "Owner"; which Project was designed by:

Robert M. Swedroe, Architect,

hereinafter called the "Architect".

NOW, THEREFORE, in consideration of the premises and of the mutual convenants and agreements hereinafter set forth, receipt and sufficiency of which are hereby mutually acknowledged, the Contractor and Subcontractor hereby agree as follows:

ARTICLE 1

THE CONTRACT DOCUMENTS

1.1 The Contract Documents for this Subcontract consist of this Agreement and any Exhibits attached hereto, the Agreement between the Owner and Contractor dated June 3, 1983, the Conditions of the Contract between the Owner and Contractor (General, Supplementary and other Conditions), all Riders and Exhibits thereto, Drawings, Specifications, all Addenda issued prior to execution of this Agreement between the Owner and Contractor, and all Modifications issued subsequent thereto. All of the above documents, which form the Contract between the Owner and Contractor, are a part of this Subcontract, are binding upon the Sucontractor, and are available for inspection by the Sucontractor. For purposes of this Agreement, reference to "the Contract Documents" shall mean and include each of the foregoing documents; reference to "this Subcontract" or "this Agreement" shall mean the within Subcontract Agreement; and reference to "the Contractor's Agreement" shall mean the aforementioned Agreement between Owner and Contractor.

133

Appendix D

1.2 The Subcontractor acknowledges that this Subcontract is subject to all of the terms and conditions of the Contractor's Agreement and the Subcontractor shall be bound to the Contractor by the terms of this Subcontract and of the Contractor's Agreement and shall assume toward the Contractor all of the obligations and responsibilities which the Contractor by the Contractor's Agreement assumes toward the Owner, insofar as applicable to this Agreement, provided that where any provision of the Contractor's Agreement is inconsistent with any provision of this Subcontract, the provisions of this Subcontract shall prevail.

ARTICLE 2

THE WORK

The Subcontractor shall furnish all labor, materials and equipment and shall perform all of the following work for (hereinafter called "the Work"):

ARTICLE 3

TIME OF COMMENCEMENT AND COMPLETION

The Subcontractor understands and acknowledges that time is of the essence of this Subcontract and all time limits stated in the Contract Documents are of the essence of this Subcontract, and any delays in performance may result in consequential damages in addition to any other damages provided for herein. No extension of time will be valid without the Contractor's written consent. The Subcontractor agrees to complete the several portions and the whole of the Work herein sublet in accordance with Contractor's schedules. Subcontractor understands and agrees that if Owner does not authorize commencement of the Project, this Subcontract shall be null and void and the parties hereto shall be released and relieved from any obligation unto each other.

ARTICLE 4

THE CONTRACT SUM

4.1 The Contractor shall pay the Subcontractor in current funds for the performance of the Work, subject to additions and deductions by Change Order, the total sum of   (Tax Included). Any amount owing at any time from the Subcontractor to Owner, contractor or any of the Subcontractor's subcontractors, materialmen, laborers or suppliers may be set off against amounts due and payable hereunder.

4.2 It is understood that at the time of the execution of this Subcontract the parties hereto are aware of the materials necessary to perform this Subcontract and/or the difficulty in obtaining same. It is accordingly understood that no claim shall be made by the Subcontractor for increase in the Contract Sum herein, even though it may be necessary to obtain materials from other warehouse stocks or otherwise, in order to perform by the Time of Completion provided herein; the Subcontractor shall at no time claim that the

-2-

Contract Sum was predicated on obtaining materials from any
particular or usual source of supply. If it be thereafter
claimed that the Subcontractor finds the price of labor and
materials herein provided for has increased to any extent
for any reason whatever, including but not limited to strikes,
force, or voluntary agreements between employer and employee,
present or future, governmental regulations, trade association
agreements, whether the same be brought about by statute,
agreement or otherwise, or any change of economic conditions
whatsoever, it is understood that any and all risks of
increase in price of labor and materials have been contemplated
by the Subcontractor and have been taken into full consideration
in arriving at the Contract Sum. The Subcontractor shall,
at no time, claim such increase by reason of delays of the
Contractor, any of the other subcontractors, the Owner or
its representatives, or other independent contractors employed
by the Owner, or for any other cause whatsoever.

ARTICLE 5

PROGRESS PAYMENTS

The Subcontractor shall submit to the Contractor at its
main office located at 1301 Dade Boulevard, Miami Beach,
Florida 33139, no later than the 25th day of each month its
requisition (Application for Payment) properly and completely
filled out on a form designated and approved by Contractor.
The Contractor shall pay the Subcontractor each progress
payment under this Subcontract within seven (7) working days
after receipt of same under its Application for Payment from
the institution providing the construction financing and/or
Owner. The amount of each progress payment to the Subcontractor
shall be equal to the percentage of the completion paid to
the Contractor for the Work under this Subcontract, applied
to the Contract Sum of this Subcontract, plus the amount
allowed for materials and equipment suitably stored by the
Subcontractor, less the aggregate previous payments to the
Subcontractor, less a ten percent (10%) retainage, and
less such other amounts which may be determined by the
Contractor to assure compliance by the Subcontractor of its
guarantee of the Work performed by the Subcontractor under
this Subcontract. It is intended by Contractor that progress
payments be paid monthly on or before the fifteenth day of
each month.

ARTICLE 6

FINAL PAYMENT

Final payment shall be due forty-five (45) after the
Contractor receives its final payment under the Contractors
Agreement which shall be paid to Contractor at such time as
Contractor's Work under the Contractor's Agreement is fully
completed and performed in accordance with the Contractor's
Agreement and has been approved and accepted by the Owner,
Architect, institution providing the construction financing,
and all other governmental agencies and authorities having
jurisdiction over the project, less such amounts as may be
determined by Contractor to assure compliance by the Subcontractor
of its guarantee of the Work performed by the Subcontractor
under this Subcontract. Before issuance of the final payment
the Subcontractor, if required by Contractor, shall submit
evidence satisfactory to the Contractor that all payrolls,
bills for material and equipment, and all indebtedness
connected with the Subcontractor's Work under this Subcontract
have been satisfied.

ARTICLE 7

PERFORMANCE AND LABOR AND MATERIAL PAYMENT BONDS

ARTICLE 8

TEMPORARY SITE FACILITIES

ARTICLE 9

INSURANCE

9.1 Prior to commencing the Work, the Subcontractor shall, at is own cost and expense, obtain all required insurance from an insurer acceptable to Contractor in amounts acceptable to the Contractor and shall furnish satisfactory evidence to Contractor of compliance with this provision. The insurance to be maintained by Subcontractor throughout the job shall include but shall not be limited to the following:

    a. Subcontractor shall maintain adequate protection of the Work, and shall protect the owners of adjacent property from injury or damage arising from the completion of the Work by Subcontractor and Subcontractor shall be responsible for any such damage or injury.

    b. The Subcontractor shall carry insurance to cover all claims under Workman's Compensation, and all other claims for damages or for personal injuries, including death which may arise from the completion of the Work by Subcontractor.

    c. The Subcontractor shall maintain Comprehensive General Including Automobile Liability Insurance, including contractual liability insurance against the liability assumed in the indemnification provisions of this Agreement as hereinafter contained and Subcontractor's Protective Liability Insurance, for the minimum amount of $500,000.00 each occurrence, $500,000.00 aggregate bodily injury and property damage combined single limit, and in addition, a $1,000,000.00 umbrella policy.

    d. The Subcontractor shall maintain a theft policy insuring Subcontractor, Owner and Contractor for all materials stolen.

9.2 The Subcontractor shall maintain the foregoing insurance in full force and effect until the final completion of the Work performed hereunder and all such policies shall name the Contractor, the Owner and any other party engaged in completing the Work as additional named insureds therein. Prior to commencing the Work, Subcontractor shall submit to Contractor certificates in duplicate evidencing such insurance to be in full force and effect affirming that all premiums have theretofore been paid, and providing that insurer notify Contractor directly of any notice of cancellation under any of said policies. Nothing contained herein shall be deemed in any way to relieve the Subcontractor from any liability under any indemnification and hold harmless agreement hereinafter contained.

9.3 If the Subcontractor shall fail to carry said insurance or to furnish Contractor with a certificate from the proper insurance company showing that Subcontractor has complied with the provisions of this Article, the Contractor may deduct and retain from any payments to become due to the Subcontractor hereunder, such amount as will indemnify it from any claims for said Workmen's Compensation Insurance premiums and Public Liability Insurance or as will be sufficient to pay premiums for said insurance, and may use said sums to pay said premiums.

ARTICLE 10

INDEMNITY BY SUBCONTRACTOR

10.1 In consideration of the first two hundred fifty dollars ($250.00) paid by the Contractor to the Subcontractor, the Subcontractor hereby agrees to indemnify and hold harmless the Owner and the Contractor and all of their agents and employees from and against any and all claims for damages, losses, suits, judgments, expenses, costs and charges of ever kind and nature, including attorney's fees, arising out of or resulting from the performance of the Subcontractor's Work under this Subcontract, whether direct or indirect, by reason of personal injuries, death or property damages sustained by any persons or others, whether such claims, losses, suits, damages, judgments, expenses and costs arise wholly or in part from acts or omissions of the Subcontractor or of any other party engaged in the performance of the Work, even though such injuries, death or damages be due to the active and affirmative negligence or breach of any nondelegable statutory duty on the part of the Contractor, its officers, agents, and employees.

10.2 In any claim and all claims against the Contractor or any of its agents or employees by any employee of the Subcontractor, anyone directly or indirectly employed by it or anyone for whose acts it may be liable, the indemnification obligation under this Article shall not be limited in any way by any limitation on the amount or type of damages, compensation or benefits payable by or for the Subcontractor under workmen's compensation acts, disability benefit acts or other employee benefit acts.

10.3 The consideration recited in this indemnity and hold harmless provision shall be applicable to any other section of this Subcontract specifically relating to indemnification of the Owner or Contractor and/or holding same harmless.

ARTICLE 11

WORKING CONDITIONS

ARTICLE 12

SUBCONTRACTOR'S RESPONSIBILITIES

12.1 The Subcontractor shall submit to the Contractor its requisition form (Application for Payment) at such times as stipulated in Article 5 to enable the Contractor to apply for payment. The requisition shall state the estimated percentage of the Work that has been satisfactorily completed.

Appendix D

12.2 If payments are made on the valuation of Work done, the Subcontractor shall, before the first application, submit to the Contractor a schedule of values of the various parts of the Work aggregating the total Contract Sum of this Subcontract made out in such detail as the Subcontractor and Contractor may agree upon, or as required by the Owner, and supported by such evidence as to its correctness as the Contractor may direct. This schedule, when approved by the Contractor, shall be used as a basis for Applications for Payment, unless it be found to be in error. In applying for payment, the Subcontractor shall submit a statement based upon this schedule.

12.3 If payments are made on account of materials or equipment not incorporated in the Work but delivered and suitably stored at the site, or at some other location agreed upon in writing, such payments shall be in accordance with the terms and conditions of the Contract Documents.

12.4 The Subcontractor shall pay for all materials or equipment and labor used in, or in connection with, the performance of the Work under this Subcontract through the period covered by previous payments received from the Contractor, and shall furnish satisfactory evidence, including but not limited to releases of lien and paid receipts, when requested by the Contractor, to verify compliance with the above requirements and that same are free from any claims, liens, or encumbrances.

12.5 The Subcontractor shall make all claims promptly to the Contractor for additional work, extensions of time, and damage for delays or otherwise, in accordance with the Contract Documents.

12.6 In carrying out the Work the Subcontractor shall take necessary precautions to protect properly the finished work of other trades from damage caused by his operations.

12.7 The Subcontractor shall at all times keep the building and premises clean of debris arising out of the Work. Unless otherwise provided, the Subcontractor shall not be held responsible for unclean conditions caused by other contractors or subcontractors.

12.8 The Subcontractor shall take all reasonable safety precautions with respect to the Work, shall comply with all safety measures initiated by the Contractor and with all applicable laws, ordinances, rules, regulations and orders of any public authority for the safety of persons or property in accordance with the requirements of the Contract Documents. The Subcontractor shall report within three days to the Contractor any injury to any of the Subcontractor's employees at the site.

12.9 The assignment by the Subcontractor of this Subcontract or any interest therein or of any money due or to become due by reason of the terms hereof without the prior written consent of the Contractor shall be void. Any assignment, if consented to in writing by the Contractor, shall be deemed subject to all of the responsibilities and liabilities of the Subcontractor, including, but not limited to, credits and set-offs and shall not relieve the Subcontractor from any liability hereunder. Any permitted assignee shall assume all obligations of the Subcontractor hereunder in writing. Further, no permitted Assignee shall release the Subcontractor from any responsibilities or liabilities hereunder.

12.10 The Subcontractor warrants that all new materials and equipment furnished and incorporated by him in the

Project shall be new unless otherwise specified, and that all Work under this Subcontract shall be of good quality, free from faults and defects and in comformance with the Contract Documents. All work not conforming to these standards may be considered defective. The warranty provided in this Paragraph 12.10 shall be in addition to and not in limitation of any other warranty or remedy required by law or by the Contract Documents.

12.11 The Subcontractor agrees that the Contractor's equipment will be available to the Subcontractor only at the Contractor's discretion and on mutually satisfactory terms.

12.12 The Subcontractor shall furnish periodic progress reports on the Work as mutually agreed, including information on the status of materials and equipment under this Subcontract which may be in the course of preparation or manufacture.

12.13 The Subcontractor shall make any and all changes in the Work from the Drawings and Specifications of the Contract Documents without invalidating this Subcontract when specifically ordered to do so in writing by the Contractor. The Subcontractor prior to the commencement of such changed or revised work, shall submit promptly to the Contractor written copies of the cost or credit proposal for such revised Work in a manner consistent with the Contract Documents.

12.14 The Subcontractor shall cooperate with the Contractor and other subcontractors whose work might interfere with the Subcontractor's Work, and shall participate in the preparation of coordinated drawings in areas of congestion as required by the Contract Documents, specifically noting and advising the Contractor of any such interference.

12.15 The Subcontractor shall cooperate with the Contractor in scheduling and performing his Work to avoid conflict or interference with the work of others.

12.16 The Subcontractor shall promptly submit shop drawings and samples as required in order to perform his work efficiently, expeditiously and in a manner that will not cause delay in the progress of the work of the Contractor or other subcontractors.

12.17 The Subcontractor shall give all notices and comply with all laws, ordinances, rules, regulations and orders of any public authority bearing on the performance of the Work under this Subcontract. The Subcontractor shall secure and pay for all permits, fees and licenses necessary for the execution of the Work described in the Contract Documents as applicable to this Subcontract. All certificates of every kind (except the certificate of occupancy) which any federal, state or local agency and/or department may issue with respect to the Work, shall be procured and delivered to the Contractor at the Subcontractor's expense immediately upon completion of the Work.

12.18. The Subcontractor shall comply with federal, state and local tax laws, social security acts, unemployment compensation acts and workmen's compensation acts insofar as applicable to the performance of this Subcontract.

12.19 The Subcontractor agrees that the Work shall be completed subject to the final approval of the Contractor.

12.20 The Subcontractor shall indemnify and hold harmless the Contractor and all of its agents and employees from and against all claims, damages, losses and expenses including attorney's fees arising out of or resulting from

Appendix D

the performance of the Subcontractor's Work under the Contract Documents.

12.21 The Subcontractor agrees upon demand to exhibit to the Contractor paid bills and releases or waivers of liens of the Subcontractor's suppliers, mechanics, laborers, and the Subcontractor's subcontractors.

12.22 If at any time there shall be evidence of any lien or claim for work or material furnished in the performance of the Work resulting from the failure of the Subcontractor to pay any of its debts or obligations which if established would be an encumbrance upon the property upon which the work is performed or for which the said Contractor might become liable, the Contractor shall have the right to retain out of any payment then due or thereafter to become due under the terms of this Subcontract, an amount sufficient to completely indemnify it against such lien or claim. Should there prove to be any such claim after all payments are made, the Subcontractor shall refund to the Contractor all monies that the latter may be compelled to pay in discharging such lien or satisfying such claim.

12.23 The Subcontractor agrees to promptly repair, fix and correct any defects in the Work and any complaints arising as a result of said Work and when requested by the Owner, Architect and/or Contractor, or its duly authorized representative. In the event the Subcontractor fails to so do, the Contractor may withold any further payments to the Subcontractor for any work done prior to or after the date when said complaint and request to repair has been transmitted to the Subcontractor. In the event the Subcontractor fails to repair, fix and correct said defects and complaints within two days after notice in writing thereof, the Contractor may proceed to do so under the provisions of Article 13.

12.24 The Subcontractor hereby waives any right it may have at law to file any lien or claim for lien against the job site or property upon which the Work is to be completed. In the event the Subcontractor files any mechanic's liens, stop orders, or otherwise attempts to encumber the premises hereinbefore described, the Contractor may, at its option, terminate this Subcontract by giving the Subcontractor two days' notice in writing of its intention to do so, without affecting any remedies or claims Contractor may have for a breach of said Subcontract or for filing said lien, stop order or encumbrance.

12.25 The Subcontractor agrees to make all necessary payments to any union or agencies to which said Subcontractor is obligated to make payment. In the event that Subcontractor shall fail to make said payments, the Contractor shall have the right to make such payment or payments, on behalf of the Subcontractor and the amount or amounts thereof shall be deducted from the Contract Sum.

12.26 The Subcontractor warrants that all materials and equipment furnished and incorporated by itself in the Project shall be new unless otherwise specified and that all Work under this Subcontract shall be of good quality, free from faults and defects and in conformance with the Contract Documents. Furthermore, all work performed and materials furnished by the Subcontractor shall be unconditionally warranted for the greater of a period of one year from the date of completion or whatever period of time which the Contractor is liable under any applicable law to any condominium unit purchaser or condominium association or to Owner under the Contract Documents. All work not conforming to these standards may be considered defective.

-8-

Appendix D   141

This warranty shall be in addition to and not in limitation of any other warranty or remedy required by law or by the Contract Documents. The Subcontractor, at its own expense, agrees to and shall immediately upon demand of the Contractor repair or replace in a manner satisfactory to the Contractor, any and all work furnished by the Subcontractor which may have become defective, due to faulty materials or workmanship, or due to unsatisfactory functioning of plant or equipment, furnished hereunder, for which the Subcontractor is at fault, within such period. The Subcontractor further agrees to pay for all damage to any structures resulting from defects in the Work and all expenses necessary to remove, replace or repair in a satisfactory manner, any other work which may be damaged or disturbed in making repairs to the Work included in this warranty.

12.27 The Subcontractor hereby represents and warrants to the Contractor that the Subcontractor: (a) is experienced and skilled in construction of the type of work provided herein; (b) has and will maintain adequate, skilled employees, work force, materials, machinery, tools and equipment necessary and adequate to accomplish the Work on or prior to the Time of Completion as provided herein; (c) has visited the jobsite and is familiar with the local conditions under which the Work is to be performed, and has correlated its observations with the requirements of this Subcontract; and (d) has carefully examined the drawings and specifications and agrees that they are adequate and suitable for the purposes intended.

12.28 In the prosecution of the Work, the Subcontractor foreman shall represent the Subcontractor and in the absence of the latter, all instructions and notifications given to the foreman shall be binding as if given to the said Subcontractor. Upon the request, however, of said foreman or of the Subcontractor, said instructions or notifications shall be reduced to writing.

12.29 Subcontractor hereby indemnifies and holds harmless the Owner and or Contractor from any and all claims, liabilities, demands, actions or causes of actions, that may be incurred or sustained by Owner and/or Contractor in connection with the Work performed by the Subcontractor under this Subcontract and in connection with Owner's and or Contractor's warranties and/or responsibilities or obligations under and pursuant to The Condominium Act, Chapter 718, Florida Statutes, as same may be amended from time to time and specifically, but without limitation, include therein §718.203, Fla.Stat. and/or any other warranties of merchantability and fitness whether arising from custom, usage, course of trade, statutory or case law, or otherwise. The foregoing indemnification includes reasonable attorney's fees and court costs (including appellate fees and court costs, if any). This provision shall be effective notwithstanding any other clause in this Subcontract and shall be deemed to be in addition to all of Owner's and or Contractor's rights under this Subcontract and not by way of limitation in any form or manner.

12.30 The Subcontractor shall give the Contractor at least ten (10) days advance notice, in writing, stating that the Work will be fully completed and ready for final inspection and/or tests.

12.31 Irrespective of whether or not a watchman is employed on the Project, during or after working hours, the Contractor shall in no event be responsible to the Subcontractor for any damage to the Work already installed by the Subcontractor or for any additional costs or expenses of Subcontractor in

142   Appendix D

repairing or replacing such damaged work, or because of any loss or damage to materials not yet installed, nor loss or damage to the Subcontractor's tools or equipment; it is the intent of the parties that liabilities and responsibilities of the Subcontractor for its own Work, shall be absolute until final acceptance of the entire Project by Owner under the Contractor's Agreement. The Subcontractor shall be required to maintain adequate insurance to protect itself against loss under this provision.

ARTICLE 13

DEFAULT BY SUBCONTRACTOR

13.1 If the Subcontractor at any time, in the opinion of the Contractor, fails, refuses or neglects in the performance of the Work to keep a competent foreman and necessary assistants and to supply a sufficient number of properly skilled workmen, or to supply and install materials of the proper quality and quantity, or fails to supply and furnish the insurance required by Article 9, or should be adjudged a bankrupt or make a general assignment for the benefit of creditors, or if a Receiver or Trustee should be appointed on account of insolvency, or if its credit should become so impaired that it is unable to pay its bills for materials purchased, or if its suppliers will not extend credit to it and refuse to ship materials, or if it cannot meet its weekly payroll expenses, or in the opinion of the Contractor, Owner, or Architect, fails in any respect to commence or prosecute the Work with sufficient promptness and diligence to insure its completion within the time designated by the Contractor, or fails in the performance of any of the agreements on its part herein contained, then in either of such events, upon giving the Subcontractor two (2) days written notice of such default or breach, the Contractor may enter upon the Project and complete the Work itself or by hiring other subcontractors the Contractor may provide such other material and workman which may be necessary to complete the Work or any part thereof, and the Contractor is hereby empowered to do so as often as it may deem necessary in order to hasten the completion of the Work without releasing the Subcontractor from liability hereunder. Further, at the Contractor's option it may, upon giving two (2) days written notice, terminate this Agreement with the right to the Contractor to complete the Work in the manner aforesaid. In any of such events, all equipment, materials, implements, appliances, or tools furnished by, or belonging to, the Subcontractor may be used by the Contractor or its agents in completing the Work, and without charge or cost to Contractor and the Contractor shall not be required to transport such equipment from the Project after such use. Contractor, in completing the Work may at its option, take over any contracts or purchase orders of the Subcontractor's which contracts and purchase orders Subcontractor does hereby assign to Contractor effective upon termination or taking over of the Work of the Subcontractor, in whole or in part, as herein provided. All notices under this Article shall be in writing and shall be delivered personally or by mail or left at the Subcontractor's place of business. All expenses of such notice or notices and of the completion of the Work and/or part thereof including costs and reasonable attorney's fees, whether related to litigation or otherwise, shall be deducted from the amount due or to become due the Subcontractor hereunder. The Subcontractor shall be liable for any excess costs of completing the Work terminated or performed by Contractor or through another party in whole or in part and shall pay the same within three (3) days after the bill is sent to the Subcontractor. The exercise of any such remedy by the Contractor shall be without prejudice to any other

remedy it may have. The remedies set forth herein are
cummulative and in addition to any other remedies available
at law or equity to Contractor.

ARTICLE 14

CONTRACTOR'S RESPONSIBILITIES

14.1 The Contractor shall be bound to the Subcontractor by the terms of this Subcontract and of the Contract Documents and shall assume toward the Subcontractor all the obligations and responsibilities that the Owner, by those Documents, assumes toward the Contractor, and Contractor shall have the benefit of all rights, remedies, and redress against the Subcontractor which the Owner, by those Documents, has against the Contractor, insofar as applicable to this Subcontract, provided that where any provision of the Contract Documents is inconsistent with any provision in this Subcontract, this Subcontract shall govern.

14.2 If there by any conflict between this Subcontract and the Contract Documents, the Subcontractor hereby agrees to be bound by the provisions which gives the greater rights to the Owner and/or Contractor.

ARTICLE 15

TERMINATION

15.1 Upon twenty-four (24) hours' notice to the Subcontractor, the Contractor reserves the right to terminate this Agreement because of restrictions by governmental authority on construction or building materials, or inability to secure materials, supplies or labor. The Subcontractor further agrees that in the event the Owner is unable to obtain construction financing for the Project or if construction financing is suspended or terminated by the lender for the Project or if the Owner terminates the Contractor's Agreement with the Contractor for any reason whatsoever including a decision by the Owner to stop construction of the Project or if the Contractor under the Contractor's Agreement terminates said agreement with Owner, then in any of such events, the Contractor may terminate this Agreement upon said twenty-four (24) hours' notice to the Subcontractor. Upon any termination of this Agreement pursuant to this Article 15, the Contractor shall not be liable to pay the Subcontractor any portion of the Contract Sum or other compensation in excess of that percentage of the Work actually completed at the time of termination. Nothing contained herein shall be in limitation of any rights the Contractor may have under any other provisions of this Subcontract. In the event Contractor should terminate this Subcontract under the provisions of this Article 15, the Subcontractor agrees that Contractor shall only be responsible to pay Subcontractor an amount equivalent to all payments received by Contractor from the Owner and/or construction lender, for the Work performed hereunder, less all payments theretofore made to Subcontractor, and less any amounts properly withheld from the Subcontractor by Contractor under this Agreement. Subcontractor agrees with Contractor that with respect to any further sums due Subcontractor under this Subcontract payment for which has not been paid and received by Contractor, it will look solely to the Owner for said payment and agrees not to make any claim against Contractor for any such payment.

ARTICLE 16

MISCELLANEOUS PROVISIONS

16.1 This Subcontract contains the entire agreement between the parties hereto. No promise, representation,

-11-

warranty or covenant not included in this Subcontract has been or is relied on by either party. Each party has relied on its own examination of the full Contract Documents and the provisions thereof, and the counsel of its own adviser. No modification or amendment to this Subcontract except as provided herein shall be of any force or effect unless in writing and executed by both parties.

16.2 This Subcontract shall inure to the benefit of and be binding upon the successors and assigns of the parties hereto.

16.3 Time is of the essence of this Subcontract. No extensions of time will be valid without the Contractor's written consent.

16.4 This Subcontract shall be governed in its enforcement and in its construction and interpretation by the laws of the State of Florida.

16.5 Each provision or portion thereof of this Subcontract shall be severable from each other provision. The invalidity of any provision or portion hereof shall not invalidate any other provision of this Subcontract.

16.6 This Subcontract may be executed in any number of counterparts which together shall constitute the agreement of the parties. The paragraph headings herein contained are for the purposes of identification only and shall not be considered in construing this Subcontract. Whenever used, the singular shall include the plural and the plural the singular and the use of any gender shall include all genders as appropriate.

16.7 Subcontractor hereby acknowledges and agrees that the Owner and the construction lender are third party beneficiaries to this Subcontract. Subcontractor further agrees that this Subcontract may be assigned by the Contractor to the Owner, or to the construction lender, and may be assumed by Owner or construction lender in the event of any termination of the Contractor's Agreement with Owner.

16.8 Should any claim of lien be filed against the Contractor, Owner or the Project, the Subcontractor shall immediately satisfy or transfer such lien to bond unless such lien resulted from the fault of the Owner, in which event the satisfaction or bonding of such lien shall be the Owner's responsibility. As a condition precedent to final payment being made to the Subcontractor, the Subcontractor shall deliver to the Contractor an affidavit in accordance with §713.06(d), Florida Statutes, stating the name of any lienor who has not been paid in full and the amount due or to become due to each such lienor for labor, services or materials furnished. The Owner and Contractor shall have the right to make direct payment to any lienor listed in said affidavit, to any lienor who has given notice to the Owner or any other lienor known to the Owner. The Subcontractor hereby specifically waives the ten (10) day notice requirement which is prerequisite to the Owner paying such bills directly to a lienor, provided that the Subcontractor has not notified the Owner of a dispute in connection with any such lienor.

16.9 Within ten (10) days after payment by the Contractor to the Subcontractor in accordance with each periodic Application for Payment, the Subcontractor shall deliver to the Contractor its partial release of lien and partial releases of lien from all sub-subcontractors and suppliers effective through the date of such payment.

IN WITNESS WHEREOF, this Subcontract has been duly executed by the parties hereto on the day and year first written above.

As to Contractor: CONTRACTOR: MILLMAN CONSTRUCTION COMPANY

_____   By:_____

_____

As to Subcontractor:

_____   SUBCONTRACTOR:

_____   By:_____

EXHIBIT "A"
(pages 1-7 inclusive)

ARCHITECTURAL DRAWINGS

Drawing #	Revision #	Latest Date	Drawing #	Revision #	Latest Date
	3	1/31/83	A-33	2	1/31/83
A-1	5	1/31/83	A-34	2	1/31/83
A-2	3	1/31/83	A-35	2	1/31/83
A-3	3	1/31/83	A-36	2	1/31/83
A-3A	3	1/31/83	A-40	2	1/31/83
A-4	3	1/31/83	A-41	2	1/31/83
A-5	3	1/31/83	A-42	2	1/31/83
A-6	3	1/31/83	A-43	2	1/31/83
A-7	3	1/31/83	A-44	1	1/31/83
A-8	3	1/31/83	A-45	2	1/31/83
A-8A	2	1/31/83	A-46	1	1/31/83
A-9	3	1/31/83	A-47	2	1/31/83
A-10	3	1/31/83	A-48	2	1/31/83
A-11	3	1/31/83	A-49	1	1/31/83
A-11A	3	1/31/83	A-50	1	1/31/83
A-12	3	1/31/83	A-51	1	1/31/83
A-13	3	1/31/83	A-52	2	1/31/83
A-13A	1	1/31/83			
A-14	3	1/31/83			
A-15	3	1/31/83			
A-16	3	1/31/83			
A-17	3	1/31/83			
A-17A	3	1/31/83			
A-18	3	1/31/83			
A-18A	3	1/31/83			
A-19	3	1/31/83			
A-19A	2	1/31/83			
A-20	3	1/31/83			
A-21	3	1/31/83			
A-21	3	1/31/83			
A-22	3	1/31/83			
A-23	3	1/31/83			
A-24	3	1/31/83			
A-25	3	1/31/83			
A-26	3	1/31/83			
A-27	3	1/31/83			
A-27A	3	1/31/83			
A-28	3	1/31/83			
A-29	3	1/31/83			
A-30	3	1/31/83			
A-31	3	1/31/83			
A-32	1	1/31/83			

Appendix D   147

EXHIBIT "A"

STRUCTURAL DRAWINGS

Drawing #	Revision #	Latest Date	Drawing #	Revision #	Latest Date
S-1	5	1/31/83	S-33	2	1/31/83
S-2	2	1/31/83	S-34	2	1/31/83
S-3	2	1/31/83	S-35	2	1/31/83
S-4	2	1/31/83	S-36	1	1/31/83
S-5	4	1/31/83	S-37	1	1/31/83
S-6	3	1/31/83	S-37A	1	1/31/83
S-7	2	1/31/83	S-38	--	1/31/83
S-8	2	1/31/83	S-39	3	1/31/83
S-9	3	1/31/83	S-40	3	1/31/83
S-10	2	1/31/83	S-41	2	1/31/83
S-11	2	1/31/83	S-42	2	1/31/83
S-12	4	1/31/83			
S-12A	2	1/31/83			
S-12B	2	1/31/83			
S-13	4	1/31/83			
S-13A	4	1/31/83			
S-14	3	1/31/83			
S-14A	3	1/31/83			
S-15	3	1/31/83			
S-16	3	1/31/83			
S-17	3	1/31/83			
S-18	3	1/31/83			
S-19	2	1/31/83			
S-19A	2	1/31/83			
S-20	3	1/31/83			
S-20A	2	1/31/83			
S-20B	2	1/31/83			
S-21	2	1/31/83			
S-22	2	1/31/83			
S-23	3	1/31/83			
S-24	5	1/31/83			
S-25	2	1/31/83			
S-26	1	1/31/83			
S-27	1	1/31/83			
S-28	3	1/31/83			
S-29	4	1/31/83			
S-30	3	1/31/83			
S-31	2	1/31/83			
S-32	2	1/31/83			

Appendix D

PLUMBING DRAWINGS

Drawing #	Revision #	Latest Date
P-1	2	1/31/83
P-2	1	1/31/83
P-3	1	1/31/83
P-4	2	1/31/83
P-5	2	1/31/83
P-6	1	1/31/83
P-7	3	1/31/83
P-7A	2	1/31/83
P-7B	1	1/31/83
P-8	2	1/31/83
P-9	2	1/31/83
P-10	1	1/31/83
P-11	--	1/31/83
P-12	4	1/31/83
P-12A	--	1/31/83
P-13	3	1/31/83
P-13A	4	1/31/83
P-13B	4	1/31/83
P-13C	2	1/31/83
P-13D	2	1/31/83
P-14	--	1/31/83
P-15	1	1/31/83
P-16	--	1/31/83
P-17	--	1/31/83

HVAC DRAWINGS

Drawing #	Revision #	Latest Date
AC-1	1	1/31/83
AC-2	--	1/31/83
AC-3	--	1/31/83
AC-4	2	1/31/83
AC-5	--	1/31/83
AC-6	1	1/31/83
AC-6A	1	1/31/83
AC-7	1	1/31/83
AC-7A	1	1/31/83
AC-7B	1	1/31/83
AC-7C	1	1/31/83
AC-8	--	1/31/83
AC-8A	2	1/31/83
AC-9	2	1/31/83
AC-10	2	1/31/83
AC-10A	2	1/31/83
AC-11	2	1/31/83
AC-12	2	1/31/83
AC-12A	1	1/31/83
AC-13	2	1/31/83
AC-14	2	1/31/83
AC-15	2	1/31/83
AC-16	2	1/31/83
AC-16A	1	1/31/83
AC-16B	2	1/31/83
AC-17	1	1/31/83
AC-19	1	1/31/83
AC-20	3	1/31/83
AC-21	2	1/31/83
AC-22	--	1/31/83

ELECTRICAL DRAWINGS

Drawing #	Revision #	Latest Date
E-1	--	1/31/83
E-1A	1	1/31/83
E-2	4	1/31/83
E-3	2	1/31/83
E-3A	3	1/31/83
E-4	2	1/31/83
E-5	1	1/31/83
E-6	1	1/31/83
E-6A	1	1/31/83
E-7	1	1/31/83
E-8	1	1/31/83
E-9	2	1/31/83
E-10	2	1/31/83
E-10A	1	1/31/83
E-11	1	1/31/83
E-12	2	1/31/83
E-13	2	1/31/83
E-14	2	1/31/83
E-14A	2	1/31/83
E-14B	2	1/31/83
E-14C	3	1/31/83
E-14D	2	1/31/83
E-14E	2	1/31/83
E-15	1	1/31/83
E-15A	--	1/31/83
E-15B	1	1/31/83
E-16	--	1/31/83
E-17	1	1/31/83
E-18	1	1/31/83
E-19	2	1/31/83
E-20	--	1/31/83
E-21	--	1/31/83
E-22	2	1/31/83
E-23	3	1/31/83
E-24	1	1/31/83
E-25	--	1/31/83
E-26	3	1/31/83
E-27	3	1/31/83
E-28	2	1/31/83
E-29	1	1/31/83
E-30	1	1/31/83
E-31	1	1/31/83
E-31A	1	1/31/83
E-32	1	1/31/83
E-33	1	1/31/83

INTERIOR DESIGN DRAWINGS

Drawing #	Revision #
ID-0	3
ID-02	3
ID-03	2
ID-04	1
ID-05	1
ID-06	1
ID-08	2
ID-09	1
ID-10	4
ID-11	2
ID-12	2
ID-13	1
ID-14	3
ID-17	2
ID-18	2
ID-18	2
ID-20	--
ID-21	3
ID-22	2
ID-23	1
ID-24	2
ID-25	--
ID-26	2
ID-27	2
ID-31	1
D-1	1
D-2	--
D-3	--
D-4	--

LANDSCAPE DRAWINGS

Drawing #	Dated
L-1	1/31/83
L-2	1/31/83
L-3	1/31/83
L-4	1/31/83
L-5	1/31/83
L-6	1/31/83
L-7	1/31/83
L-8	1/31/83
L-8A	1/31/83
L-9	1/31/83
L-10	1/31/83
L-11	1/31/83
L-12	1/31/83
L-13	1/31/83

Williams Island existing Elevations, dated 2/25/83.

SPECIFICATIONS

WILLIAMS ISLAND
TOWER "D"
INTERIOR FF&E SPECIFICATIONS PREPARED BY   DALE KELLER
DEC 1982

BUILDING SPECIFICATIONS, Dated 22 November 1982
Division 1 through 16

Addendum No. 1 dated 1/31/83
Addendum No. 2 dated 3/1/83
Addendum No. 3 dated 3/3/83
Addendum No. 4 dated 3/8/83
Addendum No. 5 dated 3/11/83

Appendix E
Requisition Printout
(General Contractor to Owner)

PROJECT:
LOCATION:
GENERAL CONTRACTOR : MILLMAN CONSTRUCTION CO., INC.

CERTIFICATE FOR PAYMENT
REQUISITION # 27
MONTH ENDING DATE: 08/31/85

DESCRIPTION OF ITEM	ORIGINAL BID AMOUNT	REVISED AMOUNT	COMPLETE TO DATE	RETAINED TO DATE	PREVIOUSLY COMPLETED	THIS ESTIMATE	BALANCE TO COMPLETE
1 GENERAL CONDITION	485,700.00		485,700.00	0.00	485,350.00	350.00	0.00
2 LAYOUT	158,000.00		158,000.00	0.00	158,000.00	0.00	0.00
3 SUPERVISION	349,000.00		349,000.00	0.00	348,000.00	1,000.00	0.00
4 HOISTING	975,323.00		975,323.00	0.00	975,323.00	0.00	0.00
5 EARTHWORK	260,150.00		260,150.00	0.00	260,150.00	0.00	0.00
6 CONCRETE	1,760,648.00		1,760,648.00	0.00	1,760,648.00	0.00	0.00
7 ROUGH CARPENTRY	129,800.00		129,800.00	0.00	129,800.00	0.00	0.00
8 MILLWORK LABOR	90,000.00		90,000.00	0.00	90,000.00	0.00	0.00
9 MILLWORK	237,780.00		237,780.00	0.00	237,780.00	0.00	0.00
10 MISC. METALS	303,282.00		303,282.00	0.00	302,282.00	1,000.00	0.00
11 FLOOR SAFES	50,000.00		50,000.00	0.00	50,000.00	0.00	0.00
12 PRECAST RAILINGS	127,600.00		127,600.00	0.00	127,600.00	0.00	0.00
13 I.D. MILLWORK	336,900.00		336,900.00	0.00	331,900.00	5,000.00	0.00
14 FORMWORK	2,198,429.00		2,198,429.00	0.00	2,198,429.00	0.00	0.00
15 PRESTRESSED SYSTEM	325,000.00		325,000.00	0.00	325,000.00	0.00	0.00
16 REINFORCING STEEL	1,410,000.00		1,410,000.00	0.00	1,410,000.00	0.00	0.00
17 MASONRY	516,000.00		516,000.00	0.00	516,000.00	0.00	0.00
18 HOLLOW METAL	97,000.00		97,000.00	0.00	97,000.00	0.00	0.00
19 PILING	4,800.00		4,800.00	0.00	4,800.00	0.00	0.00
20 APPLIANCES	740,698.00		740,698.00	0.00	740,698.00	0.00	0.00
21 ELEVATORS	685,000.00		685,000.00	0.00	685,000.00	0.00	0.00
22 SWIMMING POOLS	65,700.00		65,700.00	0.00	65,700.00	0.00	0.00
23 SAUNAS	7,224.00		7,224.00	0.00	7,224.00	0.00	0.00
24 WATERPROOFING-CRWL	125,000.00		125,000.00	0.00	125,000.00	0.00	0.00
25 WINDOWS-SLIDING GL	710,000.00		710,000.00	0.00	700,000.00	10,000.00	0.00
26 FINISHED HDWE-MED	95,000.00		95,000.00	0.00	95,000.00	0.00	0.00

DESCRIPTION OF ITEM	ORIGINAL BID AMOUNT	REVISED AMOUNT	COMPLETE TO DATE	RETAINED TO DATE	PREVIOUSLY COMPLETED	THIS ESTIMATE	BALANCE TO COMPLETE
27 MIRRORS	177,000.00		177,000.00	0.00	176,500.00	500.00	0.00
28 STUCCO	395,000.00		395,000.00	0.00	395,000.00	0.00	0.00
29 DRYWALL	1,630,000.00		1,630,000.00	0.00	1,630,000.00	0.00	0.00
30 CERAMIC TILE & MAR	989,000.00		989,000.00	0.00	989,000.00	0.00	0.00
31 SUSPENDED CEILINGS	166,000.00		166,000.00	0.00	166,000.00	0.00	0.00
32 RESILIENT FLOORING	75,000.00		75,000.00	0.00	74,500.00	500.00	0.00
33 PAINTING	584,900.00		584,900.00	0.00	577,950.00	6,950.00	0.00
34 SHOWER ENCLOSURES	60,000.00		60,000.00	0.00	59,000.00	1,000.00	0.00
35 CARPETING	180,000.00		180,000.00	0.00	180,000.00	0.00	0.00
36 TRASH CHUTES	18,000.00		18,000.00	0.00	18,000.00	0.00	0.00
37 MAILBOXES	7,500.00		7,500.00	0.00	7,500.00	0.00	0.00
38 KITCHEN CABINETS	694,000.00		694,000.00	0.00	694,000.00	0.00	0.00
39 FEATURE POOLS	42,850.00		42,850.00	0.00	42,850.00	0.00	0.00
40 ROOFING	75,300.00		75,300.00	0.00	75,300.00	0.00	0.00
41 SKYLIGHTS	75,000.00		75,000.00	0.00	75,000.00	0.00	0.00
42 AIR-CONDITIONING	1,090,000.00		1,090,000.00	0.00	1,090,000.00	0.00	0.00
43 PLUMBING	3,156,000.00		3,156,000.00	0.00	3,156,000.00	0.00	0.00
44 ELECTRICAL	2,250,000.00		2,250,000.00	0.00	2,248,500.00	1,500.00	0.00
45 INSURANCE & TAXES	337,873.00		337,873.00	0.00	337,873.00	0.00	0.00
46 PERFORMANCE BOND	112,000.00		112,000.00	0.00	112,000.00	0.00	0.00
SUB-TOTALS	24,359,457.00		24,359,457.00	0.00	24,331,657.00	27,800.00	0.00

CHANGE ORDERS	ORIGINAL BID AMOUNT	REVISED AMOUNT	COMPLETE TO DATE	RETAINED TO DATE	PREVIOUSLY COMPLETED	THIS ESTIMATE	BALANCE TO COMPLETE
001 06/17/83		79,936.00	79,936.00	0.00	79,936.00	0.00	0.00
002 06/27/83		-740,698.10	-740,698.10	0.00	-740,698.10	0.00	0.00
003 06/27/83		-180,000.00	-180,000.00	0.00	-180,000.00	0.00	0.00
004 08/26/83		-47,500.00	-47,500.00	0.00	-47,500.00	0.00	0.00
005 08/26/83		-17,100.00	-17,100.00	0.00	-17,100.00	0.00	0.00
006 08/26/83		-4,750.00	-4,750.00	0.00	-4,750.00	0.00	0.00

CHANGE ORDERS		ORIGINAL BID AMOUNT	REVISED AMOUNT	COMPLETE TO DATE	RETAINED TO DATE	PREVIOUSLY COMPLETED	THIS ESTIMATE	BALANCE TO COMPLETE
007	08/26/83		-13,775.00	-13,775.00	0.00	-13,775.00	0.00	0.00
008	08/26/83		-114,000.00	-114,000.00	0.00	-114,000.00	0.00	0.00
009	08/26/83		-17,955.00	-17,955.00	0.00	-17,955.00	0.00	0.00
010	08/26/83		-1,995.00	-1,995.00	0.00	-1,995.00	0.00	0.00
011	08/26/83		-19,950.00	-19,950.00	0.00	-19,950.00	0.00	0.00
012	08/26/83		-7,125.00	-7,125.00	0.00	-7,125.00	0.00	0.00
013	08/26/83		-4,750.00	-4,750.00	0.00	-4,750.00	0.00	0.00
014	08/26/83		-47,500.00	-47,500.00	0.00	-47,500.00	0.00	0.00
015	08/26/83		-3,800.00	-3,800.00	0.00	-3,800.00	0.00	0.00
016	09/07/83		17,100.00	17,100.00	0.00	17,100.00	0.00	0.00
017	09/07/83		-19,000.00	-19,000.00	0.00	-19,000.00	0.00	0.00
018	09/07/83		-6,175.00	-6,175.00	0.00	-6,175.00	0.00	0.00
019	09/07/83		-2,850.00	-2,850.00	0.00	-2,850.00	0.00	0.00
020	09/07/83		-4,750.00	-4,750.00	0.00	-4,750.00	0.00	0.00
021	09/07/83		-6,080.00	-6,080.00	0.00	-6,080.00	0.00	0.00
022	09/07/83		-1,710.00	-1,710.00	0.00	-1,710.00	0.00	0.00
023	09/07/83		-2,090.00	-2,090.00	0.00	-2,090.00	0.00	0.00
024	09/07/83		-1,900.00	-1,900.00	0.00	-1,900.00	0.00	0.00
025	09/07/83		-8,550.00	-8,550.00	0.00	-8,550.00	0.00	0.00
026	09/07/83		-15,456.00	-15,456.00	0.00	-15,456.00	0.00	0.00
027	09/07/83		-13,775.00	-13,775.00	0.00	-13,775.00	0.00	0.00
028	09/07/83		-28,500.00	-28,500.00	0.00	-28,500.00	0.00	0.00
029	09/07/83		-19,950.00	-19,950.00	0.00	-19,950.00	0.00	0.00
030	09/07/83		-14,250.00	-14,250.00	0.00	-14,250.00	0.00	0.00
031	09/20/83		1,437.50	1,437.50	0.00	1,437.50	0.00	0.00
032	09/23/83		28,750.00	28,750.00	0.00	28,750.00	0.00	0.00
033	10/07/83		-27,550.00	-27,550.00	0.00	-27,550.00	0.00	0.00
034	10/07/83		-8,550.00	-8,550.00	0.00	-8,550.00	0.00	0.00

CHANGE ORDERS		ORIGINAL BID AMOUNT	REVISED AMOUNT	COMPLETE TO DATE	RETAINED TO DATE	PREVIOUSLY COMPLETED	THIS ESTIMATE	BALANCE TO COMPLETE
035	10/20/83		5,291.00	5,291.00	0.00	5,291.00	0.00	0.00
036	10/20/83		7,634.00	7,634.00	0.00	7,634.00	0.00	0.00
037	11/14/83		-7,125.00	-7,125.00	0.00	-7,125.00	0.00	0.00
038	11/14/83		-4,750.00	-4,750.00	0.00	-4,750.00	0.00	0.00
039	11/14/83		-4,750.00	-4,750.00	0.00	-4,750.00	0.00	0.00
040	12/05/83		34,100.19	34,100.19	0.00	34,100.19	0.00	0.00
041	12/21/83		38,000.00	38,000.00	0.00	38,000.00	0.00	0.00
042	01/04/84		4,680.50	4,680.50	0.00	4,680.50	0.00	0.00
043	01/12/84		232,811.00	232,811.00	0.00	232,811.00	0.00	0.00
044	01/12/84		201,766.00	201,766.00	0.00	201,766.00	0.00	0.00
045	02/08/84		16,675.00	16,675.00	0.00	16,675.00	0.00	0.00
046	02/09/84		27,571.00	27,571.00	0.00	27,571.00	0.00	0.00
047	02/14/84		35,359.86	35,359.86	-0.00	35,359.86	0.00	0.00
048	02/14/84		16,675.00	16,675.00	0.00	16,675.00	0.00	0.00
049	02/20/84		17,664.00	17,664.00	0.00	17,664.00	0.00	0.00
050	02/22/84		7,601.00	7,601.00	0.00	7,601.00	0.00	0.00
051	02/29/84		32,994.00	32,994.00	0.00	32,994.00	0.00	0.00
052	03/05/84		539.06	539.06	0.00	539.06	0.00	0.00
053	03/13/84		4,623.00	4,623.00	0.00	4,623.00	0.00	0.00
054	03/16/84		4,280.76	4,280.76	-0.00	4,280.76	0.00	0.00
055	03/19/84		4,812.75	4,812.75	-0.01	4,812.75	0.00	0.00
056	03/19/84		5,000.00	5,000.00	0.00	5,000.00	0.00	0.00
057	04/10/84		8,197.20	8,197.20	0.00	8,197.20	0.00	0.00
058	04/10/84		4,315.95	4,315.95	-0.01	4,315.95	0.00	0.00
059	04/11/84		2,029.29	2,029.29	-0.00	2,029.29	0.00	0.00
060	04/19/84		10,042.11	10,042.11	0.00	10,042.11	0.00	0.00
061	04/19/84		501.40	501.40	0.00	501.40	0.00	0.00
062	04/19/84		301.88	301.88	-0.00	301.88	0.00	0.00
063	05/01/84		21,039.25	21,039.25	-0.01	21,039.25	0.00	0.00

CHANGE ORDERS		ORIGINAL BID AMOUNT	REVISED AMOUNT	COMPLETE TO DATE	RETAINED TO DATE	PREVIOUSLY COMPLETED	THIS ESTIMATE	BALANCE TO COMPLETE
064	05/04/84		1,982.60	1,982.60	0.00	1,982.60	0.00	0.00
065	05/16/84		14,443.00	14,443.00	0.00	14,443.00	0.00	0.00
066	05/16/84		507.15	507.15	-0.01	507.15	0.00	0.00
067	05/16/84		5,221.00	5,221.00	0.00	5,221.00	0.00	0.00
068	05/16/84		6,706.80	6,706.80	0.00	6,706.80	0.00	0.00
069	05/16/84		1,182.20	1,182.20	0.00	1,182.20	0.00	0.00
070	05/16/84		3,446.55	3,446.55	-0.01	3,446.55	0.00	0.00
071	05/17/84		3,060.15	3,060.15	-0.01	3,060.15	0.00	0.00
072	05/24/84		1,177.60	1,177.60	0.00	1,177.60	0.00	0.00
073	06/01/84		2,760.00	2,760.00	0.00	2,760.00	0.00	0.00
074	06/28/84		23,201.81	23,201.81	0.00	23,201.81	0.00	0.00
075	06/28/84		-19,613.10	-19,613.10	0.00	-19,613.10	0.00	0.00
076	07/13/84		7,605.00	7,605.00	0.00	7,605.00	0.00	0.00
077	07/17/84		7,400.25	7,400.25	-0.01	7,400.25	0.00	0.00
078	07/17/84		813.75	813.75	0.01	813.75	0.00	0.00
079	07/17/84		6,844.80	6,844.80	0.00	6,844.80	0.00	0.00
080	08/01/84		813.75	813.75	0.00	813.75	0.00	0.00
081	08/02/84		5,583.25	5,583.25	0.01	5,583.25	0.00	0.00
082	08/14/84		-523.18	-523.18	0.00	-523.18	0.00	0.00
083	08/21/84		2,700.20	2,700.20	0.00	2,700.20	0.00	0.00
084	08/29/84		1,552.75	1,552.75	-0.01	1,552.75	0.00	0.00
085	08/30/84		11,773.00	11,773.00	0.00	11,773.00	0.00	0.00
086	08/31/84		7,445.10	7,445.10	0.00	7,445.10	0.00	0.00
087	09/14/84		230.00	230.00	0.00	230.00	0.00	0.00
088	09/14/84		289.80	289.80	0.00	289.80	0.00	0.00
089	09/18/84		6,829.85	6,829.85	0.01	6,829.85	0.00	0.00
090	09/18/84		2,282.17	2,282.17	-0.00	2,282.17	0.00	0.00
091	10/23/84		4,134.25	4,134.25	-0.01	4,134.25	0.00	0.00

CHANGE ORDERS		ORIGINAL BID AMOUNT	REVISED AMOUNT	COMPLETE TO DATE	RETAINED TO DATE	PREVIOUSLY COMPLETED	THIS ESTIMATE	BALANCE TO COMPLETE
092	10/23/84		4,016.95	4,016.95	-0.01	4,016.95	0.00	0.00
093	10/23/84		1,835.40	1,835.40	0.00	1,835.40	0.00	0.00
094	10/23/84		862.50	862.50	0.00	862.50	0.00	0.00
095	10/23/84		1,868.75	1,868.75	-0.01	1,868.75	0.00	0.00
096	12/07/84		42,315.57	42,315.57	-0.00	38,315.57	4,000.00	0.00
097	12/07/84		3,118.92	3,118.92	0.00	3,118.92	0.00	0.00
098	12/07/84		8,459.40	8,459.40	0.00	8,459.40	0.00	0.00
099	12/07/84		945.30	945.30	0.00	945.30	0.00	0.00
100	12/21/84		13,664.30	13,664.30	0.00	13,664.30	0.00	0.00
101	12/21/84		1,380.00	1,380.00	0.00	1,380.00	0.00	0.00
102	12/21/84		2,254.00	2,254.00	0.00	2,254.00	0.00	0.00
103	12/21/84		2,640.40	2,640.40	0.00	2,640.40	0.00	0.00
104	01/24/85		3,220.00	3,220.00	0.00	3,220.00	0.00	0.00
105	01/24/85		1,937.75	1,937.75	0.00	1,937.75	0.00	0.00
106	01/24/85		4,825.20	4,825.20	0.00	4,825.20	0.00	0.00
107	01/24/85		5,451.00	5,451.00	0.00	5,451.00	0.00	0.00
108	01/24/85		59,239.00	59,239.00	0.00	57,750.00	1,489.00	0.00
109	01/24/85		31,780.25	31,780.25	-0.01	31,780.25	0.00	0.00
110	01/24/85		9,044.00	9,044.00	0.00	9,044.00	0.00	0.00
111	01/28/85		25,000.00	25,000.00	0.00	24,500.00	500.00	0.00
112	01/28/85		-15,164.00	-15,164.00	0.00	-15,164.00	0.00	0.00
113	01/28/85		12,746.25	12,746.25	0.00	12,746.25	0.00	0.00
114	02/07/85		19,895.00	19,895.00	0.00	19,895.00	0.00	0.00
115	02/14/85		3,232.00	3,232.00	0.00	3,232.00	0.00	0.00
116	02/14/85		54,308.00	54,308.00	0.00	54,308.00	0.00	0.00
117	02/18/85		7,027.65	7,027.65	-0.01	7,027.65	0.00	0.00
119	03/12/85		1,664.28	1,664.28	0.00	1,664.28	0.00	0.00
120	04/01/85		6,762.00	6,762.00	0.00	1,000.00	5,762.00	0.00
121	03/12/85		5,424.55	5,424.55	0.00	4,800.00	624.55	0.00

CHANGE ORDERS		ORIGINAL BID AMOUNT	REVISED AMOUNT	COMPLETE TO DATE	RETAINED TO DATE	PREVIOUSLY COMPLETED	THIS ESTIMATE	BALANCE TO COMPLETE
122	03/12/85		3,431.60	3,431.60	0.00	3,431.60	0.00	0.00
123	04/01/85		2,609.35	2,609.35	0.00	2,609.35	0.00	0.00
124	04/08/85		3,322.35	3,322.35	0.00	3,322.35	0.00	0.00
125	04/08/85		1,822.75	1,822.75	0.00	1,822.75	0.00	0.00
126	04/08/85		-3,800.00	-3,800.00	0.00	-3,800.00	0.00	0.00
127	04/09/85		6,122.60	6,122.60	0.00	6,122.60	0.00	0.00
128	04/12/85		1,109.75	1,109.75	0.00	1,109.75	0.00	0.00
129	05/30/85		-37,882.00	-37,882.00	0.00	-37,882.00	0.00	0.00
130	05/30/85		-3,700.00	-3,700.00	0.00	-3,700.00	0.00	0.00
131	07/01/85		617.70	617.70	0.00	0.00	617.70	0.00
SUB TOTAL		24,359,457.00						
REV. TO CONTRACT		-143,671.38						
TOTAL CONTRACT		24,215,785.62		24,215,785.62	-0.06	24,174,992.37	40,793.25	0.00
A. TOTAL DUE TO DATE		24,215,785.62						
B. RETAINED TO DATE		-0.06						
C. TO BE PD TO DATE		24,215,785.68						
D. PREVIOUSLY PAID		23,936,734.87						
E. DUE THIS PERIOD		279,050.81						

APPROVED FOR PAYMENT

OWNER: _____

BY: _____

DATE: _____

BY: _____

Appendix F
Subcontractor
Status Report Printout

RUN DATE: 25-JUL-88 PAGE 1
SUB CONTRACTOR STATUS REPORT

JOB NAME :

SUB NUMBER --- NAME ---

000020   INITIAL CONTRACT AMOUNT :   2,250,000.00

P	08/04/83	CHECK #2780	63,000.00
P	09/01/83	REDUCTION TO CONTRACT	54,000.00-
	09/09/83	CHECK #3002	20,700.00
	09/15/83	REDUCTION TO CONTRACT	45,300.00-
	10/10/83	CHECK #3136	16,353.00
	10/21/83	EXTRA TO CONTRACT	6,638.00
P	11/10/83	CK #3251	8,827.20
P	12/09/83	CHECK 3461	43,200.00
P	01/10/84	CHECK #3318	50,734.80
	01/16/84	EXTRA TO CONTRACT	73,883.00
P	02/10/84	CHECK #168	72,540.00
	02/15/84	EXTRA TO CONTRACT	30,747.70
P	02/21/84	3/10 PMT	234,403.00
	03/05/84	EXTRA TO CONTRACT	28,690.04
	03/19/84	EXTRA TO CONTRACT	395.00
P	03/23/84	4/10 PMT	110,916.00
	04/17/84	EXTRA TO CONTRACT	1,395.00
	04/17/84	EXTRA TO CONTRACT	3,753.00
P	04/24/84	5/10 PMT	267,163.40
	04/27/84	EXTRA TO CONTRACT	262.50
	04/27/84	EXTRA TO CONTRACT	386.00
	04/27/84	EXTRA TO CONTRACT	8,799.77
	04/27/84	REDUCTION TO CONTRACT	10,026.00-
	05/18/84	EXTRA TO CONTRACT	14,443.00
	05/18/84	EXTRA TO CONTRACT	4,540.00
P	05/25/84	6/10 PMT	287,028.00
P	05/29/84	7/10 PMT	219,082.50
	06/25/84	REDUCTION TO CONTRACT	21,173.00-
	06/29/84	EXTRA TO CONTRACT	5,952.00
P	07/20/84	8/10 PMT	216,874.55
	08/24/84	EXTRA TO CONTRACT	2,100.00
P	08/30/84	9/10 PMT	133,196.61
	09/15/84	EXTRA TO CONTRACT	63,450.00
P	09/24/84	EXTRA TO CONTRACT	6,474.00
	10/23/84	EXTRA TO CONTRACT	174,720.19
P	10/24/84	10/10 PMT	1,625.00
P	11/23/84	11/10 PMT	53,814.60
	11/23/84	12/10 PMT	62,330.75
	12/07/84	EXTRA TO CONTRACT	3,672.00
	12/07/84	EXTRA TO CONTRACT	2,712.10
P	12/25/84	1/10 PMT	588.00
	01/24/85	EXTRA TO CONTRACT	17,823.50
	01/25/85	EXTRA TO CONTRACT	9,860.00
	01/25/85	2/10 PMT	1,685.00
P	02/14/85	EXTRA TO CONTRACT	38,117.80
			54,685.00

RUN DATE: 25-JUL-88   SUB CONTRACTOR STATUS REPORT   PAGE 2

JOB NAME :

SUB NUMBER   --- NAME ---

```
         P    02/18/85 EXTRA TO CONTRACT                      6,111.00
         P    02/23/85 3/10 PMT                              23,471.10
         P    04/08/85 EXTRA TO CONTRACT                      1,585.00
         P    04/16/85 CK #4337 - PER MM                     34,221.82
              04/17/85 EXTRA TO CONTRACT                        960.00
         P    04/24/85 5/10 PMT                              32,759.33
         P    05/24/85 6/10 PMT                              23,719.85
         P    06/24/85 7/10 PMT                             123,240.84
         P    07/24/85 8/10 PMT                              27,538.99
         P    08/23/85 9/10 PMT                              71,519.58
              02/24/86 CREDIT ADJ PER MM                     24,000.00-
         P    02/24/86 CK #4857 - FINAL PMT                   4,617.70
```
----

59 ENTRIES FOR THIS SUB   REVISED CONTRACT AMOUNT :   2,431,917.11
                            TOTAL PAID TO DATE   :   2,431,917.11
                            BALANCE LEFT TO PAY   :   0.00

Appendix G
Job Cost Management Printout:
Budget and Actual Costs

RUN DATE: 25-JUL-88   COST CODE MASTER FILE PRINT-OUT   PAGE 6
JOB NAME :

```
        ---------COST CODE---------
   NO    DESCRIPTION
003 -101  CONC ACC - EXPANSION MATERIAL
TYPE  BUDGET      ACTUAL    PERCENT   TYPE  BUDGET   ACTUAL   PERCENT   TYPE  BUDGET   ACTUAL   PERCENT
  1   4,909     5,056.08   x102.99     6      0      0.00      0.00     11     0       0.00      0.00
  2       0         0.0                7      0      0.00      0.00     12     0       0.00      0.00
  3       0         0.00               8      0      0.00      0.00     13     0       0.00      0.00
  4       0         0.00               9      0      0.00      0.00     14     0       0.00      0.00
  5       0         0.00              10      0      0.00      0.00      *     *        *         *

003 -102  CONC ACC - VISQUEEN
TYPE  BUDGET      ACTUAL    PERCENT   TYPE  BUDGET   ACTUAL   PERCENT   TYPE  BUDGET   ACTUAL   PERCENT
  1     683     1,813.87   x265.57     6      0      0.00      0.00     11     0       0.00      0.00
  2       0         0.0                7      0      0.00      0.00     12     0       0.00      0.00
  3       0         0.00               8      0      0.00      0.00     13     0       0.00      0.00
  4   1,000       862.45    86.24      9      0      0.00      0.00     14     0       0.00      0.00
  5       0         0.00              10      0      0.00      0.00      *     *        *         *

003 -103  CONC ACC - CONCRETE SEALER
TYPE  BUDGET      ACTUAL    PERCENT   TYPE  BUDGET   ACTUAL   PERCENT   TYPE  BUDGET   ACTUAL   PERCENT
  1   8,038     1,927.00    23.98      6      0      0.00      0.00     11     0       0.00      0.00
  2       0         0.0                7      0      0.00      0.00     12     0       0.00      0.00
  3       0         0.00               8      0      0.00      0.00     13     0       0.00      0.00
  4   2,000     2,253.29   x112.66     9      0      0.00      0.00     14     0       0.00      0.00
  5       0         0.00              10      0      0.00      0.00      *     *        *         *

003 -104  CONC ACC - STYROFOAM
TYPE  BUDGET      ACTUAL    PERCENT   TYPE  BUDGET   ACTUAL   PERCENT   TYPE  BUDGET   ACTUAL   PERCENT
  1   6,387       804.62    12.60      6      0     16.53      5.51     11     0       0.00      0.00
  2       0         0.0                7    300      0.00      0.00     12     0       0.00      0.00
  3       0         0.00               8      0      0.00      0.00     13     0       0.00      0.00
  4       0         0.00               9      0      0.00      0.00     14     0       0.00      0.00
  5       0         0.00              10      0      0.00      0.00      *     *        *         *

003 -105  CONC ACC - DOVETAIL SLOTS
TYPE  BUDGET      ACTUAL    PERCENT   TYPE  BUDGET   ACTUAL   PERCENT   TYPE  BUDGET   ACTUAL   PERCENT
  1     810       654.88    80.85      6      0      0.00      0.00     11     0       0.00      0.00
  2       0         0.0                7      0      0.00      0.00     12     0       0.00      0.00
  3       0         0.00               8      0      0.00      0.00     13     0       0.00      0.00
  4       0         0.00               9      0      0.00      0.00     14     0       0.00      0.00
  5       0         0.00              10      0      0.00      0.00      *     *        *         *
```

RUN DATE: 25-JUL-88   COST CODE MASTER FILE PRINT-OUT   PAGE 7

JOB NAME :

```
----------COST CODE----------
 NO       DESCRIPTION
```

003 -106 CONC ACC - BRICK FOR REBAR
TYPE	BUDGET	ACTUAL	PERCENT	TYPE	BUDGET	ACTUAL	PERCENT	TYPE	BUDGET	ACTUAL	PERCENT
1	500	345.44	69.09	6	0	0.00		11	0	0.00	0.00
2	0	0.00		7	0	0.00		12	0	0.00	0.00
3	0	0.00		8	0	0.00		13	0	0.00	0.00
4	0	0.00		9	0	0.00		14	0	0.00	0.00
5	0	0.00		10	0	0.00					

003 -107 CONC ACC - ANTI-HYDRO
TYPE	BUDGET	ACTUAL	PERCENT	TYPE	BUDGET	ACTUAL	PERCENT	TYPE	BUDGET	ACTUAL	PERCENT
1	918	1,588.41	%173.02	6	0	0.00		11	0	0.00	0.00
2	0	0.0		7	0	0.00		12	0	0.00	0.00
3	0	0.00		8	0	0.00		13	0	0.00	0.00
4	0	0.00		9	0	298.50		14	0	0.00	0.00
5	0	0.00		10	0	0.00					

003 -108 CONC ACC - 6" WATER STOP
TYPE	BUDGET	ACTUAL	PERCENT	TYPE	BUDGET	ACTUAL	PERCENT	TYPE	BUDGET	ACTUAL	PERCENT
1	1,305	441.43	33.83	6	100	0.00		11	0	0.00	0.00
2	0	0.0		7	0	0.00		12	0	0.00	0.00
3	0	1.10		8	0	0.00		13	0	0.00	0.00
4	0	0.00		9	0	0.00		14	0	0.00	0.00
5	0	0.00		10	0	0.00					

003 -109 CONC ACC - PLAST ADD MIXTURE
TYPE	BUDGET	ACTUAL	PERCENT	TYPE	BUDGET	ACTUAL	PERCENT	TYPE	BUDGET	ACTUAL	PERCENT
1	0	0.00		6	0	0.00		11	0	0.00	0.00
2	0	0.00		7	0	0.00		12	0	0.00	0.00
3	0	0.00		8	0	0.00		13	0	0.00	0.00
4	0	0.00		9	0	0.00		14	0	0.00	0.00
5	0	0.00		10	0	0.00					

003 -110 CONC ACC - NON-METALLIC GROUT
TYPE	BUDGET	ACTUAL	PERCENT	TYPE	BUDGET	ACTUAL	PERCENT	TYPE	BUDGET	ACTUAL	PERCENT
1	390	239.09	61.31	6	0	0.00		11	0	0.00	0.00
2	0	0.0		7	0	0.00		12	0	0.00	0.00
3	0	0.00		8	0	0.00		13	0	0.00	0.00
4	100	0.00		9	0	0.00		14	0	0.00	0.00
5	0	0.00		10	0	0.00					

RUN DATE: 25-JUL-88   COST CODE MASTER FILE PRINT-OUT PAGE 8
JOB NAME :

----------COST CODE----------
 NO   DESCRIPTION

003 -111 CONC ACC - PVC SCUPPERS
TYPE BUDGET   ACTUAL   PERCENT TYPE BUDGET ACTUAL   PERCENT TYPE BUDGET ACTUAL PERCENT
 1   0   0.00   6   200   160.80   80.40   11   0   0.00
 2   0   0.00   7   0   0.00   12   0   0.00
 3   0   0.00   8   0   0.00   13   0   0.00
 4   0   0.00   9   0   0.00   14   0   0.00
 5   0   0.00   10   0   0.00   *****

003 -112 CONC ACC - KEY COVE
TYPE BUDGET   ACTUAL   PERCENT TYPE BUDGET ACTUAL   PERCENT TYPE BUDGET ACTUAL PERCENT
 1   150   663.78 X442.52   6   0   0.00   11   0   0.00
 2   0   0.0   7   0   0.00   12   0   0.00
 3   0   0.00   8   0   0.00   13   0   0.00
 4   0   0.00   9   0   0.00   14   0   0.00
 5   0   0.00   10   0   0.00   *****

003 -113 2ND FLOOR FLOATING SLAB
TYPE BUDGET   ACTUAL   PERCENT TYPE BUDGET ACTUAL   PERCENT TYPE BUDGET ACTUAL PERCENT
 1   0   185.22   6   0   383.89   11   0   0.00
 2   0   9.0   7   0   0.00   12   0   0.00
 3   0   0.00   8   0   0.00   13   0   0.00
 4   0   36.93   9   0   0.00   14   0   0.00
 5   0   148.00   10   0   0.00   *****

004 -100 CONC GARAGE - COL PADS & FTGS
TYPE BUDGET   ACTUAL   PERCENT TYPE BUDGET ACTUAL   PERCENT TYPE BUDGET ACTUAL PERCENT
 1   1,056   0.00   6   0   0.00   11   0   0.00
 2   0   1,111.5   X105.25   7   0   0.00   12   0   0.00
 3   0   0.00   8   0   0.00   13   0   0.00
 4   1,090 1,191.36 X109.29   9   0   0.00   14   0   0.00
 5   0   0.00   10   0   0.00   *****

004 -102 CONC GARAGE - VERTS PAD TO 2
TYPE BUDGET   ACTUAL   PERCENT TYPE BUDGET ACTUAL   PERCENT TYPE BUDGET ACTUAL PERCENT
 1   0   0.00   6   0   0.00   11   0   0.00
 2   147   118.5   80.61   7   0   0.00   12   0   0.00
 3   0   0.00   8   0   0.00   13   0   0.00
 4   775   756.45   97.61   9   0   0.00   14   0   0.00
 5   0   0.00   10   0   0.00   *****

RUN DATE: 25-JUL-88   COST CODE MASTER FILE PRINT-OUT   PAGE   9
JOB NAME :
-------COST CODE-------
   NO   DESCRIPTION
004 -103 CONC GARAGE - VERTS 2 TO 3

TYPE	BUDGET	ACTUAL	PERCENT	TYPE	BUDGET	ACTUAL	PERCENT
1	0	0.00		6	0	0.00	
2	60	75.5	%125.83	7	0	0.00	
3	0	0.00		8	0	0.00	
4	436	614.93	%141.03	9	0	0.00	
5	0	0.00		10	0	0.00	

004 -104 CONC GARAGE - VERTS 3 TO P/D

TYPE	BUDGET	ACTUAL	PERCENT	TYPE	BUDGET	ACTUAL	PERCENT
1	0	0.00		6	0	0.00	
2	60	83.5	%139.16	7	0	0.00	
3	0	0.00		8	0	0.00	
4	436	801.20	%183.76	9	0	0.00	
5	0	0.00		10	0	0.00	

004 -201 CONC GARAGE - SLAB 1ST LEVEL

TYPE	BUDGET	ACTUAL	PERCENT	TYPE	BUDGET	ACTUAL	PERCENT
1	0	0.00		6	0	0.00	
2	516	636.5	%123.35	7	0	0.00	
3	0	0.00		8	0	0.00	
4	1,453	1,896.49	%130.52	9	0	0.00	
5	3,126	3,770.50	%120.61	10	0	0.00	

004 -202 CONC GARAGE - SLAB 2ND LEVEL NORTH HALF (NO PSI)

TYPE	BUDGET	ACTUAL	PERCENT	TYPE	BUDGET	ACTUAL	PERCENT
1	0	0.00		6	0	0.00	
2	666	679.5	%102.02	7	0	0.00	
3	0	0.00		8	0	0.00	
4	1,744	1,503.30	86.20	9	0	0.00	
5	1,714	1,914.25	%111.68	10	0	0.00	

004 -203 CONC GARAGE - SLAB 2ND LEVEL (PSI)

TYPE	BUDGET	ACTUAL	PERCENT	TYPE	BUDGET	ACTUAL	PERCENT
1	0	0.00		5	0	0.00	
2	240	420.0	%175	7	0	0.00	
3	0	0.00		8	0	0.00	
4	872	1,583.15	%181.55	9	0	0.00	
5	2,285	2,792.18	%122.19	10	0	0.00	

RUN DATE: 25-JUL-88   COST CODE MASTER FILE PRINT-OUT   PAGE 10

JOB NAME :

----------------COST CODE----------------
NO   DESCRIPTION

004 -204 CONC GARAGE - SLAB 3RD LEVEL
TYPE   BUDGET   ACTUAL   PERCENT   TYPE   BUDGET   ACTUAL   PERCENT   TYPE   BUDGET   ACTUAL   PERCENT
 1   0   0.00   0.00   6   0   0.00   0.00   *11   0   0.00   0.00
 2   276   385.0   X139.49   7   0   0.00   0.00   *12   0   0.00   0.00
 3   0   0.00   0.00   * 8   0   0.00   0.00   *13   0   0.00   0.00
 4   872   1,211.77   X138.96   * 9   0   0.00   0.00   *14   0   0.00   0.00
 5   2,285   1,518.35   66.45   *10   0   0.00   0.00

004 -205 CONC GARAGE - SLAB POOL DECK
TYPE   BUDGET   ACTUAL   PERCENT   TYPE   BUDGET   ACTUAL   PERCENT   TYPE   BUDGET   ACTUAL   PERCENT
 1   0   0.00   0.00   6   0   0.00   0.00   *11   0   0.00   0.00
 2   355   631.0   X177.74   * 7   0   0.00   0.00   *12   0   0.00   0.00
 3   0   0.00   0.00   * 8   0   0.00   0.00   *13   0   0.00   0.00
 4   1,163   2,583.35   X222.12   * 9   0   0.00   0.00   *14   0   0.00   0.00
 5   2,285   3,298.25   X144.34   *10   0   0.00   0.00

004 -206 CONC GARAGE - SLAB POOL PAVILLION ROOF   ************ *NEW ACTIVITY* ************
TYPE   BUDGET   ACTUAL   PERCENT   TYPE   BUDGET   ACTUAL   PERCENT   TYPE   BUDGET   ACTUAL   PERCENT
 1   0   0.00   0.00   6   0   0.00   0.00   *11   0   0.00   0.00
 2   13   25.0   X192.30   * 7   0   0.00   0.00   *12   0   0.00   0.00
 3   0   0.00   0.00   * 8   0   0.00   0.00   *13   0   0.00   0.00
 4   36   147.72   X410.33   * 9   0   0.00   0.00   *14   0   0.00   0.00
 5   143   0.00   0.00   *10   0   0.00   0.00

004 -207 CONC GARAGE - TOPPING 2ND LEVL
TYPE   BUDGET   ACTUAL   PERCENT   TYPE   BUDGET   ACTUAL   PERCENT   TYPE   BUDGET   ACTUAL   PERCENT
 1   0   0.00   0.00   6   0   0.00   0.00   *11   0   0.00   0.00
 2   239   271.0   X113.38   * 7   0   0.00   0.00   *12   0   0.00   0.00
 3   0   0.00   0.00   * 8   0   0.00   0.00   *13   0   0.00   0.00
 4   654   1,292.05   X197.56   * 9   0   0.00   0.00   *14   0   0.00   0.00
 5   1,714   1,952.70   X113.92   *10   0   0.00   0.00

004 -208 CONC GARAGE - TOPPING POOL DK
TYPE   BUDGET   ACTUAL   PERCENT   TYPE   BUDGET   ACTUAL   PERCENT   TYPE   BUDGET   ACTUAL   PERCENT
 1   0   0.00   0.00   6   0   0.00   0.00   *11   0   0.00   0.00
 2   396   431.5   X108.96   * 7   0   0.00   0.00   *12   0   0.00   0.00
 3   0   0.00   0.00   * 8   0   0.00   0.00   *13   0   0.00   0.00
 4   872   2,362.93   X270.97   * 9   0   0.00   0.00   *14   0   0.00   0.00
 5   2,285   1,994.91   87.30   *10   0   0.00   0.00

RUN DATE: 25-JUL-88   COST CODE MASTER FILE PRINT-OUT   PAGE 11

JOB NAME :

----------COST CODE----------
 NO   DESCRIPTION

004 -209 CONC GARAGE - SLAB POOL & WHPL
TYPE BUDGET   ACTUAL   PERCENT   TYPE BUDGET   ACTUAL PERCENT   TYPE BUDGET   ACTUAL PERCENT
 1   0   0.00   6   0   0.00   11   0   0.00
 2   64   138.0   %215.62   7   0   0.00   12   0   0.00
 3   0   0.00   8   0   0.00   13   0   0.00
 4   218   627.72   %287.95   9   0   0.00   14   0   0.00
 5   286   343.50   %120.10   10   0   0.00   * * * * *

004 -302 CONC GARAGE - BEAMS 2ND LEVEL
TYPE BUDGET   ACTUAL   PERCENT   TYPE BUDGET   ACTUAL PERCENT   TYPE BUDGET   ACTUAL PERCENT
 1   0   0.00   6   0   0.00   11   0   0.00
 2   227   0.0   7   0   0.00   12   0   0.00
 3   0   0.00   8   0   0.00   13   0   0.00
 4   1,550   0.00   9   0   0.00   14   0   0.00
 5   0   0.00   10   0   0.00   * * * * *

004 -303 CONC GARAGE - BEAMS 3RD LEVEL
TYPE BUDGET   ACTUAL   PERCENT   TYPE BUDGET   ACTUAL PERCENT   TYPE BUDGET   ACTUAL PERCENT
 1   0   0.00   6   0   0.00   11   0   0.00
 2   176   0.00   7   0   0.00   12   0   0.00
 3   0   0.00   8   0   0.00   13   0   0.00
 4   1,162   0.00   9   0   0.00   14   0   0.00
 5   0   0.00   10   0   0.00   * * * * *

004 -304 CONC GARAGE - BEAMS POOL DECK
TYPE BUDGET   ACTUAL   PERCENT   TYPE BUDGET   ACTUAL PERCENT   TYPE BUDGET   ACTUAL PERCENT
 1   0   0.00   6   0   0.00   11   0   0.00
 2   344   0.0   7   0   0.00   12   0   0.00
 3   0   0.00   8   0   0.00   13   0   0.00
 4   1,938   0.00   9   0   0.00   14   0   0.00
 5   0   0.00   10   0   0.00   * * * * *

004 -401 CONC GARAGE - WALLS 1ST LEVEL
TYPE BUDGET   ACTUAL   PERCENT   TYPE BUDGET   ACTUAL PERCENT   TYPE BUDGET   ACTUAL PERCENT
 1   0   0.00   6   0   0.00   11   0   0.00
 2   241   241.0   %100   7   0   0.00   12   0   0.00
 3   0   0.00   8   0   0.00   13   0   0.00
 4   775   904.46   %116.70   9   0   0.00   14   0   9.03
 5   0   0.00   10   0   0.00   * * * * *

RUN DATE: 25-JUL-88   COST CODE MASTER FILE PRINT-OUT PAGE 12
JOB NAME :

----------COST CODE----------
 NO   DESCRIPTION

004 -402 CONC GARAGE - WALLS 2ND LEVEL
TYPE BUDGET   ACTUAL   PERCENT   TYPE BUDGET ACTUAL PERCENT   TYPE BUDGET ACTUAL PERCENT
 1   0   0.00   6   0   0.00   *11   0   0.00
 2   121   95.0   78.51   7   0   0.00   *12   0   0.00
 3   0   0.00   8   0   0.00   *13   0   0.00
 4   582   747.92   *128.50   9   0   0.00   *14   0   0.00
 5   0   0.00   10   0   0.00   *

004 -403 CONC GARAGE - WALLS 3RD LEVEL
TYPE BUDGET   ACTUAL   PERCENT   TYPE BUDGET ACTUAL PERCENT   TYPE BUDGET ACTUAL PERCENT
 1   0   0.00   6   0   0.00   *11   0   0.00
 2   76   23.0   30.26   7   0   0.00   *12   0   0.00
 3   0   0.00   8   0   0.00   *13   0   0.00
 4   388   158.99   40.98   9   0   0.00   *14   0   0.00
 5   0   0.00   10   0   0.00   *

004 -404 CONC GARAGE - WALLS POOL DECK
TYPE BUDGET   ACTUAL   PERCENT   TYPE BUDGET ACTUAL PERCENT   TYPE BUDGET ACTUAL PERCENT
 1   0   0.00   6   0   0.00   *11   0   0.00
 2   324   556.5   *171.75   7   0   0.00   *12   0   0.00
 3   0   0.00   8   0   0.00   *13   0   0.00
 4   1,453   3,737.15 *257.20   9   0   0.00   *14   0   0.00
 5   0   0.00   10   0   0.00   *

004 -405 CONC GARAGE - WALLS POOL & WHP
TYPE BUDGET   ACTUAL   PERCENT   TYPE BUDGET ACTUAL PERCENT   TYPE BUDGET ACTUAL PERCENT
 1   0   0.00   6   0   0.00   *11   0   0.00
 2   90   80.0   88.89   7   0   0.00   *12   0   0.00
 3   0   0.00   8   0   0.00   *13   0   0.00
 4   388   446.08 *114.96   9   0   0.00   *14   0   0.00
 5   0   0.00   10   0   0.00   *

004 -406 SLAB IN WATER FEATURE
TYPE BUDGET   ACTUAL   PERCENT   TYPE BUDGET ACTUAL PERCENT   TYPE BUDGET ACTUAL PERCENT
 1   0   0.0   6   0   0.00   *11   0   0.00
 2   0   0.0   7   0   0.00   *12   0   0.00
 3   0   0.00   8   0   0.00   *13   0   0.00
 4   0   98.48   9   0   0.00   *14   0   0.00
 5   0   184.00   10   0   0.00   *

RUN DATE: 25-JUL-88   COST CODE MASTER FILE PRINT-OUT PAGE 13

JOB NAME :

------COST CODE------
NO   DESCRIPTION

005 -101 CONC MISC - TIE COLUMNS
TYPE	BUDGET	ACTUAL	PERCENT	TYPE	BUDGET	ACTUAL	PERCENT
1	0	0.00		* 6	0	0.00	
2	155	152.0	98.06	* 7	0	0.00	
3	0	0.00		* 8	0	0.00	
4	1,550	3,831.37	%247.18	* 9	0	0.00	
5	0	55.50		* 10	0	0.00	

TYPE	BUDGET	ACTUAL	PERCENT
* 11	0	0.00	
* 12	0	0.00	
* 13	0	0.00	
* 14	0	0.00	

005 -102 CONC MISC - CELL BLOCK FILLED IN TRANSFORMER VAULT
TYPE	BUDGET	ACTUAL	PERCENT	TYPE	BUDGET	ACTUAL	PERCENT
1	0	0.00		* 6	0	0.00	
2	31	20.5	66.13	* 7	0	0.00	
3	0	0.00		* 8	0	0.00	
4	310	791.92	%255.45	* 9	0	0.00	
5	0	0.00		* 10	0	0.00	

TYPE	BUDGET	ACTUAL	PERCENT
* 11	0	0.00	
* 12	0	0.00	
* 13	0	0.00	
* 14	0	0.00	

005 -103 CONC MISC - 6" WALL IN TRANS-FORMER VAULT
TYPE	BUDGET	ACTUAL	PERCENT	TYPE	BUDGET	ACTUAL	PERCENT
1	0	0.00		* 6	0	0.00	
2	12	6.0	50.00	* 7	0	0.00	
3	0	0.00		* 8	0	0.00	
4	120	135.41	%112.84	* 9	0	0.00	
5	0	0.00		* 10	0	0.00	

TYPE	BUDGET	ACTUAL	PERCENT
* 11	0	0.00	
* 12	0	0.00	
* 13	0	0.00	
* 14	0	0.00	

005 -104 CONC MISC - FORMED CONC PAVERS
TYPE	BUDGET	ACTUAL	PERCENT	TYPE	BUDGET	ACTUAL	PERCENT
1	0	0.00		* 6	0	0.00	
2	5	0.0		* 7	0	0.00	
3	0	0.00		* 8	0	0.00	
4	50	0.00		* 9	0	0.00	
5	0	0.00		* 10	0	0.00	

TYPE	BUDGET	ACTUAL	PERCENT
* 11	0	0.00	
* 12	0	0.00	
* 13	0	0.00	
* 14	0	0.00	

005 -105 CONC MISC - FILL CELL BLOCK AROUND TRASH CHUTE
TYPE	BUDGET	ACTUAL	PERCENT	TYPE	BUDGET	ACTUAL	PERCENT
1	0	0.00		* 6	0	0.00	
2	43	0.0		* 7	0	0.00	
3	0	0.00		* 8	0	0.02	
4	430	0.00		* 9	0	2.00	
5	0	0.00		* 10	0	0.00	

TYPE	BUDGET	ACTUAL	PERCENT
* 11	0	0.00	
* 12	0	0.00	
* 13	0	0.00	
* 14	0	0.00	

RUN DATE: 25-JUL-88   COST CODE MASTER FILE PRINT-OUT   PAGE 14
JOB NAME :

---COST CODE---
NO   DESCRIPTION

005-106 CONC MISC - FILL CELL BLOCK INPLANTERS ON PENTHOUSE ROOF
TYPE	BUDGET	ACTUAL	PERCENT	TYPE	BUDGET	ACTUAL	PERCENT	TYPE	BUDGET	ACTUAL	PERCENT
1	0	0.00		* 6	0	0.00		* 11	0	0.00	
2	22	68.0	X309.09	* 7	0	0.00		* 12	0	0.00	
3	0	0.00		* 8	0	0.00		* 13	0	0.00	
4	220	2,353.85	X1069.5	* 9	0	0.00		* 14	0	0.00	
5		127.73		* 10	0	0.00					

005-107 CONC MISC - FILL CELL BLOCK ATTRASH ROOM - GROUND FLOOR
TYPE	BUDGET	ACTUAL	PERCENT	TYPE	BUDGET	ACTUAL	PERCENT	TYPE	BUDGET	ACTUAL	PERCENT
1	0	0.00		* 6	0	0.00		* 11	0	0.00	
2	12	4.0	33.33	* 7	0	0.00		* 12	0	0.00	
3	0	0.00		* 8	0	0.00		* 13	0	0.00	
4	120	106.89	89.08	* 9	0	0.00		* 14	0	0.00	
5		0.00		* 10	0	0.00					

005-108 LOBBY CHANGE
TYPE	BUDGET	ACTUAL	PERCENT	TYPE	BUDGET	ACTUAL	PERCENT	TYPE	BUDGET	ACTUAL	PERCENT
1	0	16.28		* 6	0	0.00		* 11	0	0.00	
2	0	0.0		* 7	0	0.00		* 12	0	0.00	
3	0	168.84		* 8	0	0.00		* 13	0	0.00	
4	0	466.24		* 9	0	0.00		* 14	0	0.00	
5		0.00		* 10	0	0.00					

006-101 CONC SITE & POOL DECK - WALKWAY ON POOL DECK
TYPE	BUDGET	ACTUAL	PERCENT	TYPE	BUDGET	ACTUAL	PERCENT	TYPE	BUDGET	ACTUAL	PERCENT
1	0	0.00		* 6	0	0.00		* 11	0	0.00	
2	88	49.0	55.68	* 7	0	0.00		* 12	0	0.00	
3	0	0.00		* 8	0	0.00		* 13	0	0.00	
4	484	469.37	96.98	* 9	0	0.00		* 14	0	0.00	
5	571	147.20	25.78	* 10	0	0.00					

006-102 CONC SITE & POOL DECK - STEPS ON POOL DECK
TYPE	BUDGET	ACTUAL	PERCENT	TYPE	BUDGET	ACTUAL	PERCENT	TYPE	BUDGET	ACTUAL	PERCENT
1	0	0.00		* 6	0	0.00		* 11	0	0.00	
2	14	22.0	X157.14	* 7	0	0.00		* 12	0	0.00	
3	0	0.00		* 8	0	0.00		* 13	0	0.00	
4	140	172.34	X123.1	* 9	0	0.00		* 14	0	0.00	
5	96	259.53	X270.33	* 10	0	0.00					

```
RUN DATE: 25-JUL-88        COST CODE MASTER FILE PRINT-OUT PAGE 15

JOB NAME :

-------------COST CODE-------------
  NO     DESCRIPTION

006 -103   CONC SITE & POOL DECK            CONCRETE BANDS ON POOL DECK
TYPE  BUDGET      ACTUAL      PERCENT     TYPE  BUDGET      ACTUAL      PERCENT     TYPE  BUDGET      ACTUAL      PERCENT

 1       0         0.00                     6       0         0.00                  *11       0         0.00
 2      11        44.0       %400           7       0         0.00                  *12       0         0.00
 3       0         0.00                     8       0         0.00                  *13       0         0.00
 4     110       289.29      %262.98        9       0         0.00                  *14       0         0.00
 5     175       623.45      %356.25       10       0         0.00

006 -104   CONC SITE & POOL DECK            SITE FOOTINGS & GRADE BEAMS
TYPE  BUDGET      ACTUAL      PERCENT     TYPE  BUDGET      ACTUAL      PERCENT     TYPE  BUDGET      ACTUAL      PERCENT

 1       0         0.00                     6       0         0.00                  *11       0         0.00
 2     152        63.00       41.45         7       0         0.00                  *12       0         0.00
 3       0         0.00                     8       0         0.00                  *13       0         0.00
 4     582       282.35       48.51         9       0         0.00                  *14       0         0.00
 5       0         0.00                    10       0         0.00

006 -105   CONC SITE & POOL DECK            SITE SLABS
TYPE  BUDGET      ACTUAL      PERCENT     TYPE  BUDGET      ACTUAL      PERCENT     TYPE  BUDGET      ACTUAL      PERCENT

 1       0         0.00                     6       0         0.00                  *11       0         0.00
 2      27        21.0        77.78         7       0         0.00                  *12       0         0.00
 3       0         0.00                     8       0         0.00                  *13       0         0.00
 4      97       240.25      %247.46        9       0         0.00                  *14       0         0.00
 5     143       184.00      %128.67       10       0         0.00

006 -106   CONC SITE & POOL DECK            SITE WALLS
TYPE  BUDGET      ACTUAL      PERCENT     TYPE  BUDGET      ACTUAL      PERCENT     TYPE  BUDGET      ACTUAL      PERCENT

 1       0         0.00                     6       0         0.00                  *11       0         0.00
 2     270       171.0        63.33         7       0         0.00                  *12       0         0.00
 3       0         0.00                     8       0         0.00                  *13       0         0.00
 4   1,356     1,127.93       83.18         9       0         0.00                  *14       0         0.00
 5       0       332.20                    10       0         0.00

006 -107   CONC SITE & POOL DECK            SITE CONCRETE BANDS
TYPE  BUDGET      ACTUAL      PERCENT     TYPE  BUDGET      ACTUAL      PERCENT     TYPE  BUDGET      ACTUAL      PERCENT

 1       0         0.00                     6       0         0.00                  *11       0         0.00
 2      55         0.00                     7       0         0.00                  *12       0         0.00
 3       0         0.00                     8       0         0.00                  *13       0         0.00
 4     194         0.00                     9       0         0.00                  *14       0         0.00
 5     286         0.00                    10       0         0.00
```

RUN DATE: 25-JUL-88   COST CODE MASTER FILE PRINT-OUT PAGE 16

JOB NAME :

------- COST CODE -------
NO   DESCRIPTION

006 -108 CONC SITE & POOL DECK   SITE SIDEWALKS W/ BROOM FINISH

TYPE	BUDGET	ACTUAL	PERCENT	TYPE	BUDGET	ACTUAL	PERCENT	TYPE	BUDGET	ACTUAL	PERCENT
1	0	0.00		6	0	0.00		11	0	0.00	
2	98	1.0	1.02	7	0	0.00		12	0	0.00	
3	0	0.00		8	0	0.00		13	0	0.00	
4	484	31.53	6.51	9	0	0.00		14	0	0.00	
5	857	0.00		10	0	0.00					

006 -109 CONC SITE & POOL DECK - SITE SIDEWALKS W/SPECIAL FINISH

TYPE	BUDGET	ACTUAL	PERCENT	TYPE	BUDGET	ACTUAL	PERCENT	TYPE	BUDGET	ACTUAL	PERCENT
1	0	0.00		6	0	0.00		11	0	0.00	
2	0	0.00		7	0	0.00		12	0	0.00	
3	0	0.00		8	0	0.00		13	0	0.00	
4	0	0.00		9	0	0.00		14	0	0.00	
5	0	0.00		10	0	0.00					

006 -110 CONC SITE & POOL DECK   SITE CURB   *······*NEW ACTIVITY*······*

TYPE	BUDGET	ACTUAL	PERCENT	TYPE	BUDGET	ACTUAL	PERCENT	TYPE	BUDGET	ACTUAL	PERCENT
1	0	0.00		6	0	0.00		11	0	0.00	
2	20	4.0	20.00	7	0	0.00		12	0	0.00	
3	0	0.00		8	0	0.00		13	0	0.20	
4	200	21.02	10.51	9	0	0.00		14	0	0.00	
5	150	0.00		10	0	0.00					

006 -111 CONC SITE & POOL DECK - POOL DECK - 4" SIDEWALK W/CHAT

TYPE	BUDGET	ACTUAL	PERCENT	TYPE	BUDGET	ACTUAL	PERCENT	TYPE	BUDGET	ACTUAL	PERCENT
1	0	0.00		6	0	0.00		11	0	0.00	
2	42	2.0	4.76	7	0	0.00		12	0	0.00	
3	0	0.00		8	0	0.00		13	0	0.00	
4	420	36.93	8.79	9	0	0.00		14	0	0.00	
5	130	0.00		10	0	0.00					

007 -099 CONC TOWER - MUD SLAB

TYPE	BUDGET	ACTUAL	PERCENT	TYPE	BUDGET	ACTUAL	PERCENT	TYPE	BUDGET	ACTUAL	PERCENT
1	0	0.00		6	0	0.00		11	0	0.00	
2	140	126.0	90.00	7	0	0.00		12	0	0.00	
3	0	0.00		8	0	0.00		13	0	0.00	
4	1,065	845.69	79.41	9	0	0.00		14	0	0.00	
5	510	0.00		10	0	0.00					

RUN DATE: 25-JUL-88   COST CODE MASTER FILE PRINT-OUT PAGE 17

JOB NAME :

------COST CODE------

NO	DESCRIPTION									

007 -100   CONC TOWER - RAFT SLAB

TYPE	BUDGET	ACTUAL	PERCENT	TYPE	BUDGET	ACTUAL	PERCENT	TYPE	BUDGET	ACTUAL	PERCENT
1	0	0.00		6	0	0.00		11	0	0.00	
2	3,354	3,346.0	99.76	7	0	0.00		12	0	0.00	
3	0	0.00		8	0	0.00		13	0	0.00	
4	3,758	2,400.83	63.89	9	0	0.00		14	0	0.00	
5	1,494	752.88	50.39	10	0	0.00		*	0	0.00	

007 -101   CONC TOWER - PADS & FOOTINGS

TYPE	BUDGET	ACTUAL	PERCENT	TYPE	BUDGET	ACTUAL	PERCENT	TYPE	BUDGET	ACTUAL	PERCENT
1	0	0.00		6	0	0.00		11	0	0.00	
2	251	215.5	85.86	7	0	0.00		12	0	0.00	
3	0	0.00		8	0	0.00		13	0	0.00	
4	494	452.05	91.51	9	0	0.00		14	0	0.00	
5	0	0.00		10	0	0.00		*	0	0.00	

007 -102   CONC TOWER - VERTS - MAT TO 2

TYPE	BUDGET	ACTUAL	PERCENT	TYPE	BUDGET	ACTUAL	PERCENT	TYPE	BUDGET	ACTUAL	PERCENT
1	0	0.00		6	0	0.00		11	0	0.00	
2	36	352.0	97.51	7	0	0.00		12	0	0.00	
3	0	0.00		8	0	0.00		13	0	0.00	
4	1,529	2,329.27	*152.33	9	0	0.00		14	0	0.00	
5	0	0.00		10	0	0.00		*	0	0.00	

007 -103   CONC TOWER - VERTS - 2 TO 3

TYPE	BUDGET	ACTUAL	PERCENT	TYPE	BUDGET	ACTUAL	PERCENT	TYPE	BUDGET	ACTUAL	PERCENT
1	0	0.00		6	0	0.00		11	0	0.00	
2	240	265.0	*110.41	7	0	0.00		12	0	0.00	
3	0	100.00		8	0	0.00		13	0	0.00	
4	1,164	1,549.68	*133.13	9	0	0.00		14	0	0.00	
5	0	0.00		10	0	0.00		*	0	0.00	

007 -104   CONC TOWER - VERTS - 3 TO 4

TYPE	BUDGET	ACTUAL	PERCENT	TYPE	BUDGET	ACTUAL	PERCENT	TYPE	BUDGET	ACTUAL	PERCENT
1	0	0.00		6	0	0.00		11	0	0.00	
2	266	232.0	87.22	7	0	0.00		12	0	0.00	
3	0	100.00		8	0	0.00		13	0	0.00	
4	1,646	1,518.11	92.23	9	0	0.00		14	0	0.00	
5	0	0.00		10	0	0.00		*	0	0.00	

RUN DATE: 25-JUL-88   COST CODE MASTER FILE PRINT-OUT PAGE 18

JOB NAME :

----------COST CODE----------
NO   DESCRIPTION

007 -105 CONC TOWER - VERTS - 4 TO 5
TYPE	BUDGET	ACTUAL	PERCENT	TYPE	BUDGET	ACTUAL	PERCENT
1	0	0.00		6	0	0.00	
2	132	269.0	%203.78	* * 7	0	0.00	
3	0	0.00		* 8	0	0.00	
4	751	1,033.32	%137.59	* * 9	0	0.00	
5	0	0.00		* * 10	0	0.00	

007 -106 CONC TOWER - VERTS - 5 TO 6
TYPE	BUDGET	ACTUAL	PERCENT	TYPE	BUDGET	ACTUAL	PERCENT
1	0	0.00		6	0	0.00	
2	132	128.5	97.35	* * 7	0	0.00	
3	0	0.00		* 8	0	0.00	
4	753	1,332.14	%176.91	* * 9	0	0.00	
5	0	0.00		* * 10	0	0.00	

007 -107 CONC TOWER - VERTS - 6 TO 7
TYPE	BUDGET	ACTUAL	PERCENT	TYPE	BUDGET	ACTUAL	PERCENT
1	0	0.00		6	0	0.00	
2	132	124.5	94.32	* * 7	0	0.00	
3	0	0.00		* 8	0	0.00	
4	753	888.92	%118.05	* * 9	0	0.00	
5	0	0.00		* * 10	0	0.00	

007 -108 VERTICALS 7 TO 8
TYPE	BUDGET	ACTUAL	PERCENT	TYPE	BUDGET	ACTUAL	PERCENT
1	0	8.45		6	0	0.00	
2	132	123.0	93.18	* * 7	0	0.00	
3	0	0.00		* 8	0	0.00	
4	753	730.71	97.04	* * 9	0	0.00	
5	0	0.00		* * 10	0	0.00	

007 -109 CONC TOWER - VERTS - 8 TO 9
TYPE	BUDGET	ACTUAL	PERCENT	TYPE	BUDGET	ACTUAL	PERCENT
1	0	0.00		6	0	0.00	
2	132	144.0	%109.09	* * 7	0	0.00	
3	0	0.00		* 8	0	0.00	
4	753	898.35	%119.30	* * 9	0	0.00	
5	0	0.00		* * 10	0	0.00	

RUN DATE: 25-JUL-88   COST CODE MASTER FILE PRINT-OUT PAGE 19

JOB NAME :

------------COST CODE------------
 NO   DESCRIPTION

007 -110 CONC TOWER - VERTS - 9 TO 10

TYPE	BUDGET	ACTUAL	PERCENT	TYPE	BUDGET	ACTUAL	PERCENT
1	0	0.00		6	0	0.00	
2	132	127.0	96.21	7	0	0.00	
3	0	0.00		8	0	0.00	
4	753	822.42	*109.21	9	0	0.00	
5	0	0.00		10	0	0.00	

TYPE	BUDGET	ACTUAL	PERCENT
*11	0	0.00	
*12	0	0.00	
*13	0	0.00	
*14	0	0.00	

007 -111 CONC TOWER - VERTS - 10 TO 11

TYPE	BUDGET	ACTUAL	PERCENT	TYPE	BUDGET	ACTUAL	PERCENT
1	0	0.00		6	0	0.00	
2	132	126.0	95.45	7	0	0.00	
3	0	0.00		8	0	0.00	
4	753	1,016.12	*134.94	9	0	0.00	
5	0	0.00		10	0	0.00	

TYPE	BUDGET	ACTUAL	PERCENT
*11	0	0.00	
*12	0	0.00	
*13	0	0.00	
*14	0	0.00	

007 -112 CONC TOWER - VERTS - 11 TO 12

TYPE	BUDGET	ACTUAL	PERCENT	TYPE	BUDGET	ACTUAL	PERCENT
1	0	0.00		6	0	0.00	
2	132	135.0	*102.27	7	0	0.00	
3	0	0.00		8	0	0.00	
4	753	,4.61	*126.90	9	0	0.00	
5	0	0.00		10	0	0.00	

TYPE	BUDGET	ACTUAL	PERCENT
*11	0	0.00	
*12	0	0.00	
*13	0	0.00	
*14	0	0.00	

007 -114 CONC TOWER - VERTS - 12 TO 14

TYPE	BUDGET	ACTUAL	PERCENT	TYPE	BUDGET	ACTUAL	PERCENT
1	0	0.00		6	0	0.00	
2	132	123.0	93.18	7	0	0.00	
3	0	0.00		8	0	0.00	
4	753	841.99	*111.81	9	0	0.00	
5	0	0.00		10	0	0.00	

TYPE	BUDGET	ACTUAL	PERCENT
*11	0	0.00	
*12	0	0.00	
*13	0	0.00	
*14	0	0.00	

007 -115 CONC TOWER - VERTS - 14 TO 15

TYPE	BUDGET	ACTUAL	PERCENT	TYPE	BUDGET	ACTUAL	PERCENT
1	0	0.00		6	0	0.00	
2	132	120.0	90.91	7	0	0.00	
3	0	0.00		8	0	0.00	
4	753	839.02	*111.42	9	0	0.00	
5	0	0.00		10	0	0.00	

TYPE	BUDGET	ACTUAL	PERCENT
*11	0	0.00	
*12	0	0.00	
*13	0	0.00	
*14	0	0.00	

RUN DATE: 25-JUL-88   COST CODE MASTER FILE PRINT-OUT PAGE 20

JOB NAME :

---------------COST CODE---------------
NO   DESCRIPTION

007 -116 CONC TOWER - VERTS - 15 TO 16
TYPE	BUDGET	ACTUAL	PERCENT	TYPE	BUDGET	ACTUAL	PERCENT
1	0	0.00		* 11	0	0.00	
2	132	107.0	81.06	* 12	0	0.00	
3	0	0.00	x108.00	* 13	0	0.00	
4	753	813.28		* 14	0	0.00	
5	0	0.00		*			

007 -117 CONC TOWER - VERTS - 16 TO 17
TYPE	BUDGET	ACTUAL	PERCENT	TYPE	BUDGET	ACTUAL	PERCENT
1	0	0.00		* 11	0	0.00	
2	132	113.5	85.98	* 12	0	0.00	
3	0	0.00	x121.19	* 13	0	0.00	
4	753	912.57		* 14	0	0.00	
5	0	0.00		*			

007 -118 CONC TOWER - VERTS - 17 TO 18
TYPE	BUDGET	ACTUAL	PERCENT	TYPE	BUDGET	ACTUAL	PERCENT
1	0	0.00		* 11	0	0.00	
2	132	116.0	87.88	* 12	0	0.00	
3	0	0.00	x104.30	* 13	0	0.00	
4	753	785.42		* 14	0	0.00	
5	0	0.00		*			

007 -119 CONC TOWER-VERTS- 18-19
TYPE	BUDGET	ACTUAL	PERCENT	TYPE	BUDGET	ACTUAL	PERCENT
1	0	0.00		* 11	0	0.00	
2	132	117.0	88.64	* 12	0	0.00	
3	0	0.00	x148.92	* 13	0	0.00	
4	753	1,121.38		* 14	0	0.00	
5	0	0.00		*			

007 -120 CONC TOWER - VERTS - 19 TO 20
TYPE	BUDGET	ACTUAL	PERCENT	TYPE	BUDGET	ACTUAL	PERCENT
1	0	0.00		* 11	0	0.00	
2	132	105.0	79.55	* 12	0	0.00	
3	0	0.00	x122.91	* 13	0	0.00	
4	753	925.58		* 14	0	0.00	
5	0	0.00		*			

RUN DATE: 25-JUL-88   COST CODE MASTER FILE PRINT-OUT PAGE 21

JOB NAME :

```
----------COST CODE----------
NO      DESCRIPTION
```

007 -121 CONC TOWER - VERTS - 20 TO 21
TYPE	BUDGET	ACTUAL	PERCENT	TYPE	BUDGET	ACTUAL	PERCENT
1	0	0.00		* 6	0	0.00	
2	132	111.0	84.09	* 7	0	0.00	
3	0	0.00		* 8	0	0.00	
4	753	795.38	x105.62	* 9	0	0.00	
5	0	0.00		* 10	0	0.00	
				* 11			
				* 12			
				* 13			
				* 14			

007 -122 CONC TOWER - VERTS - 21 TO 22
TYPE	BUDGET	ACTUAL	PERCENT	TYPE	BUDGET	ACTUAL	PERCENT
1	0	0.00		* 6	0	0.00	
2	132	124.0	93.94	* 7	0	0.00	
3	0	0.00		* 8	0	0.00	
4	753	759.59	x100.87	* 9	0	0.00	
5	0	0.00		* 10	0	0.00	
				* 11			
				* 12			
				* 13			
				* 14			

007 -123 CONC TOWER - VERTS - 22 TO 23
TYPE	BUDGET	ACTUAL	PERCENT	TYPE	BUDGET	ACTUAL	PERCENT
1	0	0.00		* 6	0	0.00	
2	132	136.0	x103.03	* 7	0	0.00	
3	0	0.00		* 8	0	0.00	
4	753	822.59	x109.24	* 9	0	0.00	
5	0	0.00		* 10	0	0.00	
				* 11			
				* 12			
				* 13			
				* 14			

007 -124 CONC TOWER - VERTS - 23 TO 24
TYPE	BUDGET	ACTUAL	PERCENT	TYPE	BUDGET	ACTUAL	PERCENT
1	0	0.00		* 6	0	0.00	
2	132	111.0	84.09	* 7	0	0.00	
3	0	0.00		* 8	0	0.00	
4	753	584.14	77.57	* 9	0	0.00	
5	0	0.00		* 10	0	0.00	
				* 11			
				* 12			
				* 13			
				* 14			

007 -125 CONC TOWER - VERTS - 24 TO 25
TYPE	BUDGET	ACTUAL	PERCENT	TYPE	BUDGET	ACTUAL	PERCENT
1	0	0.00		* 6	0	0.00	
2	132	121.0	91.67	* 7	0	0.00	
3	0	0.00		* 8	0	0.00	
4	753	745.91	99.06	* 9	0	0.00	
5	0	0.00		* 10	0	0.00	
				* 11			
				* 12			
				* 13			
				* 14			

RUN DATE: 25-JUL-88   COST CODE MASTER FILE PRINT-OUT PAGE 22

JOB NAME :

------------------COST CODE------------------
NO   DESCRIPTION

007 -126 CONC TOWER - VERTS 25-26
TYPE	BUDGET	ACTUAL	PERCENT	TYPE	BUDGET	ACTUAL	PERCENT
1	0	0.00		6	0	0.00	0.00
2	132	103.0	78.03	7	0	0.00	0.00
3	0	0.00		8	0	0.00	0.00
4	753	662.08	87.93	9	0	0.00	0.00
5	0	0.00		10	0	0.00	

007 -127 VERTICALS 26-27
TYPE	BUDGET	ACTUAL	PERCENT	TYPE	BUDGET	ACTUAL	PERCENT
1	0	0.00		6	0	0.00	0.00
2	132	93.5	70.83	7	0	0.00	0.00
3	0	0.00		8	0	0.00	0.00
4	753	819.75	%108.86	9	0	0.00	0.00
5	0	0.00		10	0	0.00	

007 -128 CONC TOWER - VERTS - 27 TO 28
TYPE	BUDGET	ACTUAL	PERCENT	TYPE	BUDGET	ACTUAL	PERCENT
1	0	0.00		6	0	0.00	0.00
2	132	121.0	91.67	7	0	0.00	0.00
3	0	0.00		8	0	0.00	0.00
4	753	560.12	74.39	9	0	0.00	0.00
5	0	0.00		10	0	0.00	

007 -129 CONC TOWER - VERTS - 28 TO 29
TYPE	BUDGET	ACTUAL	PERCENT	TYPE	BUDGET	ACTUAL	PERCENT
1	0	0.00		6	0	0.00	0.00
2	132	120.0	90.91	7	0	0.00	0.00
3	0	0.00		8	0	0.00	0.00
4	753	901.30	%119.69	9	0	0.00	0.00
5	0	0.00		10	0	0.00	

007 -130 CONC TOWER - VERTS - 29 TO TOWER SUITE
TYPE	BUDGET	ACTUAL	PERCENT	TYPE	BUDGET	ACTUAL	PERCENT
1	0	0.00		6	0	0.00	0.00
2	132	120.0	90.91	7	0	0.00	0.00
3	0	0.00		8	0	0.00	0.00
4	753	553.68	73.53	9	0	0.00	0.00
5	0	0.00		10	0	0.00	

RUN DATE: 25-JUL-88   COST CODE MASTER FILE PRINT-OUT PAGE 23

JOB NAME :

----------COST CODE----------
 NO   DESCRIPTION

007 -131 CONC TOWER - VERTICALS - TOWERSUITE TO P/H (LOWER LEVEL)
TYPE BUDGET   ACTUAL   PERCENT   TYPE   BUDGET   ACTUAL   PERCENT
 1   0   0.00   6   0   0.00
 2   98   86.0   87.76   7   0   0.00
 3   0   0.00   8   0   0.00
 4   482   501.17   %103.97   9   0   0.00
 5   0   0.00   10   0   0.00

007 -132 CONC TOWER - VERTS - P/H LOWERLEVEL TO P/H UPPER LEVEL
TYPE BUDGET   ACTUAL   PERCENT   TYPE   BUDGET   ACTUAL   PERCENT
 1   0   0.00   6   0   0.00
 2   106   91.0   85.85   7   0   0.00
 3   0   0.00   8   0   0.00
 4   682   605.85   88.83   9   0   0.00
 5   0   0.00   10   0   0.00

007 -133 CONC TOWER - VERTS - P/H UPPERLEVEL TO ROOF & MACHINE ROOM
TYPE BUDGET   ACTUAL   PERCENT   TYPE   BUDGET   ACTUAL   PERCENT
 1   0   0.00   11   0   0.00
 2   88   114.0   %129.54   12   0   0.00
 3   0   0.00   13   0   0.00
 4   482   906.26   %188.02   14   0   0.00
 5   0   0.00

007 -134 CONC TOWER - VERTS - TO HIGH ROOF
TYPE BUDGET   ACTUAL   PERCENT   TYPE   BUDGET   ACTUAL   PERCENT
 1   0   0.00   11   0   0.00
 2   35   48.0   %137.14   12   0   0.00
 3   0   0.00   13   0   0.00
 4   282   526.91   %186.84   14   0   0.00
 5   0   0.00

007 -135 CONC TOWER - VERTICALS - TO PARAPET
TYPE BUDGET   ACTUAL   PERCENT   TYPE   BUDGET   ACTUAL   PERCENT
 1   0   0.00   11   0   0.00
 2   8   17.0   %212.5   12   0   0.00
 3   0   0.00   13   0   0.00
 4   94   181.65   %193.24   14   0   0.00
 5   0   0.00

RUN DATE: 25-JUL-88   COST CODE MASTER FILE PRINT-OUT   PAGE 24

JOB NAME :

---------------COST CODE---------------
  NO   DESCRIPTION

007 -201 CONC TOWER - SLABS - GROUND FL
TYPE BUDGET   ACTUAL   PERCENT TYPE BUDGET ACTUAL PERCENT TYPE BUDGET ACTUAL PERCENT
1   0   0.00   6   0   0.00   *11   0   0.00
2   295   392.5   %133.05 7   0   0.00   *12   0   0.00
3   0   0.00   *8   0   0.00   *13   0   0.00
4   706   1,648.35   %233.47 *9   0   0.00   *14   0   0.00
5   1,368   2,123.48   %155.22 *10

007 -202 CONC TOWER - SLABS - 2ND FLOOR
TYPE BUDGET   ACTUAL   PERCENT TYPE BUDGET ACTUAL PERCENT TYPE BUDGET ACTUAL PERCENT
1   0   0.00   6   0   0.00   *11   0   0.00
2   661   698.5   %105.67 7   0   0.00   *12   0   0.00
3   0   0.00   *8   0   0.00   *13   0   0.00
4   1,552   1,958.58   %126.19 *9   0   0.00   *14   0   0.00
5   2,052   2,926.55   %142.61 *10

007 -203 CONC TOWER - SLABS - 3RD FLOOR
TYPE BUDGET   ACTUAL   PERCENT TYPE BUDGET ACTUAL PERCENT TYPE BUDGET ACTUAL PERCENT
1   0   0.00   6   0   0.00   *11   0   0.00
2   611   493.0   80.69   7   0   0.00   *12   0   0.00
3   0   0.00   *8   0   0.00   *13   0   0.00
4   1,693   1,965.30   %116.08 *9   0   0.00   *14   0   0.00
5   2,052   3,016.55   %147.00 *10

007 -204 CONC TOWER - SLABS - 4TH FLOOR
TYPE BUDGET   ACTUAL   PERCENT TYPE BUDGET ACTUAL PERCENT TYPE BUDGET ACTUAL PERCENT
1   0   0.00   6   0   0.00   *11   0   0.00
2   836   887.0   %106.1   7   0   0.00   *12   0   0.00
3   0   0.00   *8   0   0.00   *13   0   0.00
4   2,246   2,417.96   %107.65 *9   0   0.00   *14   0   0.00
5   2,736   2,771.90   %101.31 *10

007 -205 CONC TOWER - SLABS - 5TH FLOOR
TYPE BUDGET   ACTUAL   PERCENT TYPE BUDGET ACTUAL PERCENT TYPE BUDGET ACTUAL PERCENT
1   0   0.00   6   0   0.00   *11   0   0.00
2   416   412.0   99.04   7   0   0.00   *12   0   0.00
3   0   0.00   *8   0   0.00   *13   0   0.00
4   847   1,130.97   %133.52 *9   0   0.00   *14   0   0.00
5   1,739   1,807.80   %103.95 *10

RUN DATE: 25-JUL-88　　　COST CODE MASTER FILE PRINT-OUT　PAGE 25
JOB NAME :

----------COST CODE----------
NO　　DESCRIPTION

007 -206　CONC TOWER - SLABS - 6TH FLOOR

TYPE	BUDGET	ACTUAL	PERCENT	TYPE	BUDGET	ACTUAL	PERCENT
1	0	0.00	*	6	0	0.00	0.00
2	416	424.0	%101.92	7	0	0.00	0.00
3	0	0.00	*	8	0	0.00	0.00
4	847	1,170.14	%138.15	9	0	0.00	0.00
5	1,739	1,824.00	%104.88	10	0	0.00	0.00
				11	0	0.00	*
				12	0	0.00	*
				13	0	0.00	*
				14	0	0.00	*

007 -207　CONC TOWER - SLABS - 7TH FLOOR

TYPE	BUDGET	ACTUAL	PERCENT	TYPE	BUDGET	ACTUAL	PERCENT
1	0	0.00	*	6	0	0.00	0.00
2	416	419.0	%100.72	7	0	0.00	0.00
3	0	0.00	*	8	0	0.00	0.00
4	847	932.17	%110.05	9	0	0.00	0.00
5	1,739	1,830.53	%105.26	10	0	0.00	0.00
				11	0	0.00	*
				12	0	0.00	*
				13	0	0.00	*
				14	0	0.00	*

007 -208　CONC TOWER - SLABS - 8TH FLOOR

TYPE	BUDGET	ACTUAL	PERCENT	TYPE	BUDGET	ACTUAL	PERCENT
1	0	0.00	*	6	0	0.00	0.00
2	416	416.0	%100	7	0	0.00	0.00
3	0	0.00	*	8	0	0.00	0.00
4	847	948.97	%112.03	9	0	0.00	0.00
5	1,739	1,726.43	99.28	10	0	0.00	0.00
				11	0	0.00	*
				12	0	0.00	*
				13	0	0.00	*
				14	0	0.00	*

007 -209　CONC TOWER - SLABS - 9TH FLOOR

TYPE	BUDGET	ACTUAL	PERCENT	TYPE	BUDGET	ACTUAL	PERCENT
1	0	0.00	*	6	0	0.00	0.00
2	416	410.0	98.56	7	0	0.00	0.00
3	0	0.00	*	8	0	0.00	0.00
4	847	803.52	94.87	9	0	0.00	0.00
5	1,739	1,538.80	88.49	10	0	0.00	0.00
				11	0	0.00	*
				12	0	0.00	*
				13	0	0.00	*
				14	0	0.00	*

007 -210　CONC TOWER - SLABS - 10TH FLR

TYPE	BUDGET	ACTUAL	PERCENT	TYPE	BUDGET	ACTUAL	PERCENT
1	0	0.00	*	6	0	0.00	0.00
2	416	429.0	%103.12	7	0	0.00	0.00
3	0	0.00	*	8	0	0.00	0.00
4	847	932.02	%110.03	9	0	0.00	0.00
5	1,739	1,799.55	%103.48	10	0	0.00	0.00
				11	0	0.00	*
				12	0	0.00	*
				13	0	0.00	*
				14	0	0.00	*

```
RUN DATE: 25-JUL-88       COST CODE MASTER FILE PRINT-OUT PAGE 26
JOB NAME :
          ----------COST CODE----------
    NO       DESCRIPTION
```

007 -211 CONC TOWER - SLABS - 11TH FLR
TYPE	BUDGET	ACTUAL	PERCENT	TYPE	BUDGET	ACTUAL	PERCENT	TYPE	BUDGET	ACTUAL	PERCENT
1	0	0.00		6	0	0.00		11	0	0.00	
2	416	410.0	98.56	7	0	0.00	*	12	0	0.00	
3	0	0.00	*	8	0	0.00	*	13	0	0.00	
4	847	1,177.78	%139.05	9	0	0.00	*	14	0	0.00	
5	1,739	1,711.15	98.40	10	0	0.00	*				

007 -212 CONC TOWER - SLABS - 12TH FLR
TYPE	BUDGET	ACTUAL	PERCENT	TYPE	BUDGET	ACTUAL	PERCENT	TYPE	BUDGET	ACTUAL	PERCENT
1	0	18.06		6	0	0.00		11	0	0.00	
2	416	420.0	%100.96	7	0	0.00	*	12	0	0.00	
3	0	0.00	*	8	0	0.00	*	13	0	0.00	
4	847	1,066.70	%125.93	9	0	0.00	*	14	0	0.00	
5	1,739	1,793.03	%103.10	10	0	0.00	*				

007 -214 CONC TOWER - SLABS - 14TH FLR
TYPE	BUDGET	ACTUAL	PERCENT	TYPE	BUDGET	ACTUAL	PERCENT	TYPE	BUDGET	ACTUAL	PERCENT
1	0	27.09		6	0	0.00		11	0	0.00	
2	416	409.0	98.32	7	0	0.00	*	12	0	0.00	
3	0	0.00	*	8	0	0.00	*	13	0	0.00	
4	847	1,198.62	%141.51	9	0	0.00	*	14	0	0.00	
5	1,739	1,990.83	%114.48	10	0	0.00	*				

007 -215 CONC TOWER - SLABS - 15TH FLR
TYPE	BUDGET	ACTUAL	PERCENT	TYPE	BUDGET	ACTUAL	PERCENT	TYPE	BUDGET	ACTUAL	PERCENT
1	0	0.00		6	0	0.00		11	0	0.00	
2	416	430.0	%103.36	7	0	0.00	*	12	0	0.00	
3	0	0.00	*	8	0	0.00	*	13	0	0.00	
4	847	1,014.49	%119.77	9	0	0.00	*	14	0	0.00	
5	1,739	1,775.25	%102.08	10	0	0.00	*				

007 -216 CONC TOWER - SLABS - 16TH FLR
TYPE	BUDGET	ACTUAL	PERCENT	TYPE	BUDGET	ACTUAL	PERCENT	TYPE	BUDGET	ACTUAL	PERCENT
1	0	0.00		6	0	0.00		11	0	0.00	
2	416	436.0	%104.80	7	0	0.00	*	12	0	0.00	
3	0	0.00	*	8	0	0.00	*	13	0	0.00	
4	847	1,019.66	%120.38	9	0	0.00	*	14	0	0.00	
5	1,739	2,346.70	%134.94	10	0	0.00	*				

RUN DATE: 25-JUL-88   COST CODE MASTER FILE PRINT-OUT   PAGE 27

JOB NAME :

----------COST CODE----------
NO   DESCRIPTION

007 -217 CONC TOWER - SLABS - 17TH FLR
TYPE	BUDGET	ACTUAL	PERCENT	TYPE	BUDGET	ACTUAL	PERCENT	TYPE	BUDGET	ACTUAL	PERCENT
1	0	0.00		6	0	0.00		11	0	0.00	
2	416	476.0	%114.42	7	0	0.00		12	0	0.00	
3	0	0.00		8	0	0.00		13	0	0.00	
4	847	990.75	%116.97	9	0	0.00		14	0	0.00	
5	1,739	2,082.33	%119.74	10	0	0.00		*	*	*	*

007 -218 CONC TOWER - SLABS - 18TH FLR
TYPE	BUDGET	ACTUAL	PERCENT	TYPE	BUDGET	ACTUAL	PERCENT	TYPE	BUDGET	ACTUAL	PERCENT
1	0	0.00		6	0	0.00		11	0	0.00	
2	416	427.0	%102.64	7	0	0.00		12	0	0.00	
3	0	0.00		8	0	0.00		13	0	0.00	
4	847	972.65	%114.83	9	0	0.00		14	0	0.00	
5	1,739	1,842.85	%105.97	10	0	0.00		*	*	*	*

007 -219 CONC TOWER - SLABS - 19TH FLR
TYPE	BUDGET	ACTUAL	PERCENT	TYPE	BUDGET	ACTUAL	PERCENT	TYPE	BUDGET	ACTUAL	PERCENT
1	0	0.00		6	0	0.00		11	0	0.00	
2	416	417.0	%100.24	7	0	0.00		12	0	0.00	
3	0	0.00		8	0	0.00		13	0	0.00	
4	847	1,019.22	%120.33	9	0	0.00		14	0	0.00	
5	1,739	1,844.65	%106.07	10	0	0.00		*	*	*	*

007 -220 CONC TOWER - SLABS - 20TH FLR
TYPE	BUDGET	ACTUAL	PERCENT	TYPE	BUDGET	ACTUAL	PERCENT	TYPE	BUDGET	ACTUAL	PERCENT
1	0	0.00		6	0	0.00		11	0	0.00	
2	416	432.0	%103.84	7	0	0.00		12	0	0.00	
3	0	0.00		8	0	0.00		13	0	0.00	
4	847	1,076.57	%127.10	9	0	0.00		14	0	0.00	
5	1,739	1,807.80	%103.95	10	0	0.00		*	*	*	*

007 -221 CONC TOWER - SLABS - 21ST FLR
TYPE	BUDGET	ACTUAL	PERCENT	TYPE	BUDGET	ACTUAL	PERCENT	TYPE	BUDGET	ACTUAL	PERCENT
1	0	0.00		6	0	0.00		11	0	0.00	
2	416	447.0	%107.45	7	0	0.00		12	0	0.00	
3	0	0.00		8	0	0.00		13	0	0.00	
4	847	1,199.00	%141.55	9	0	0.00		14	0	0.00	
5	1,739	1,954.28	%112.37	10	0	0.00		*	*	*	*

RUN DATE: 25-JUL-86   COST CODE MASTER FILE PRINT-OUT PAGE 28

JOB NAME :

```
----------COST CODE----------
 NO     DESCRIPTION
```

007 -222 CONC TOWER - SLABS - 22ND FLR
TYPE	BUDGET	ACTUAL	PERCENT	TYPE	BUDGET	ACTUAL	PERCENT
1	0	0.00		6 * * * *			
2	416	423.0	%101.68	7 * * * *			
3	0	0.00		8 * * * *			
4	847	882.03	%104.13	9	0	0.00	
5	1,739	1,726.43	99.28	10	0	0.00	
				11	0	0.00	
				12	0	0.00	
				13	0	0.00	
				14	0	0.00	

007 -223 CONC TOWER - SLABS - 23RD FLR
TYPE	BUDGET	ACTUAL	PERCENT	TYPE	BUDGET	ACTUAL	PERCENT
1	0	0.00		6 * * * *			
2	416	425.0	%102.16	7 * * * *			
3	0	0.00		8 * * * *			
4	847	890.36	%105.11	9	0	0.00	
5	1,739	1,938.00	%111.44	10	0	0.00	
				11	0	0.00	
				12	0	0.00	
				13	0	0.00	
				14	0	0.00	

007 -224 CONC TOWER SLABS 24 FLOOR
TYPE	BUDGET	ACTUAL	PERCENT	TYPE	BUDGET	ACTUAL	PERCENT
1	0	0.00		6 * * * *			
2	416	427.0	%102.64	7 * * * *			
3	0	0.00		8 * * * *			
4	847	845.74	99.85	9	0	0.00	
5	1,739	1,901.15	%109.32	10	0	0.00	
				11	0	0.00	
				12	0	0.00	
				13	0	0.00	
				14	0	0.00	

007 -225 CONC TOWER - SLABS - 25TH FLR
TYPE	BUDGET	ACTUAL	PERCENT	TYPE	BUDGET	ACTUAL	PERCENT
1	0	0.00		6 * * * *			
2	416	430.0	%103.36	7 * * * *			
3	0	0.00		8 * * * *			
4	847	890.33	%105.11	9	0	0.00	
5	1,739	2,087.70	%120.05	10	0	0.00	
				11	0	0.00	
				12	0	0.00	
				13	0	0.00	
				14	0	0.00	

007 -226 CONC TOWER - SLABS - 26TH FLR
TYPE	BUDGET	ACTUAL	PERCENT	TYPE	BUDGET	ACTUAL	PERCENT
1	0	0.00		6 * * * *			
2	416	439.0	%105.52	7 * * * *			
3	0	0.00		8 * * * *			
4	847	876.62	%103.49	9	0	0.00	
5	1,739	1,638.75	94.24	10	0	0.00	
				11	0	0.00	
				12	0	0.00	
				13	0	0.00	
				14	0	0.00	

```
RUN DATE: 25-JUL-88        COST CODE MASTER FILE PRINT-OUT  PAGE 29
JOB NAME :
     ----------COST CODE----------
     NO         DESCRIPTION
     ------------------------------
007 -227  CONC TOWER - SLABS - 27TH FLR
TYPE      BUDGET      ACTUAL      PERCENT      TYPE      BUDGET      ACTUAL      PERCENT
 1           0          0.00                     6          0          0.00        0.00
 2         416        433.0      %104.08         7          0          0.00        0.00
 3           0          0.00                     8          0          0.00        0.00
 4         847        987.84     %116.62         9          0          0.00        0.00
 5       1,739      1,842.85     %105.97        10          0          0.00        0.00

007 -228  CONC TOWER - SLABS - 28TH FLR
TYPE      BUDGET      ACTUAL      PERCENT      TYPE      BUDGET      ACTUAL      PERCENT
 1           0          0.00                     6          0          0.00        0.00
 2         416        423.0      %101.68         7          0          0.00        0.00
 3           0          0.00                     8          0          0.00        0.00
 4         847        997.13     %117.72         9          0          0.00        0.00
 5       1,739      2,142.90     %123.22        10          0          0.00        0.00

007 -229  CONC TOWER SLABS 29TH FLOOR
TYPE      BUDGET      ACTUAL      PERCENT      TYPE      BUDGET      ACTUAL      PERCENT
 1           0          0.00                     6          0          0.00        0.00
 2         416        424.0      %101.92         7          0          0.00        0.00
 3           0          0.00                     8          0          0.00        0.00
 4         847        841.31      99.33          9          0          0.00        0.00
 5       1,739      1,845.65     %106.13        10          0          0.00        0.00

007 -230  CONC TOWER - SLABS - TOWER SUITE
TYPE      BUDGET      ACTUAL      PERCENT      TYPE      BUDGET      ACTUAL      PERCENT
 1           0          0.00                     6          0          0.00        0.00
 2         425        425.0      %100            7          0          0.00        0.00
 3           0          0.00                     8          0          0.00        0.00
 4         847        931.95     %110.03         9          0          0.00        0.00
 5       1,739      1,424.55      81.92         10          0          0.00        0.00

007 -231  CONC TOWER - SLABS - P/H LOWERLEVEL
TYPE      BUDGET      ACTUAL      PERCENT      TYPE      BUDGET      ACTUAL      PERCENT
 1           0        110.25                     6          0          0.00        0.00
 2         433        443.0      %102.30         7          0          0.00        0.00
 3           0          0.00                     8          0          0.00        0.00
 4         988      1,428.98     %144.63         9          0          0.00        0.00
 5       1,739      2,890.60     %166.22        10          0          0.00        0.00
```

RUN DATE: 25-JUL-88   COST CODE MASTER FILE PRINT-OUT PAGE 30
JOB NAME :

------------COST CODE------------
     NO   DESCRIPTION

007 -232 CONC TOWER - SLABS - P/H UPPER LEVEL
TYPE BUDGET   ACTUAL   PERCENT   TYPE   BUDGET   ACTUAL   PERCENT   TYPE   BUDGET   ACTUAL PERCENT
 1   0   0.00   6   0   11   0   0.00
 2   709   774.0   %109.16   7   0   12   0   0.00
 3   0   0.00   8   0   13   0   0.00
 4   1,623   2,407.04   %148.30   9   0   14   0   0.00
 5   1,835   3,728.69   %203.19   10   0

007 -233 CONC TOWER - SLABS - P/H ROOF
TYPE BUDGET   ACTUAL   PERCENT   TYPE   BUDGET   ACTUAL   PERCENT   TYPE   BUDGET   ACTUAL PERCENT
 1   0   0.00   6   0   24.22   11   0   0.00
 2   332   385.0   %115.96   7   0   0.00   12   0   0.00
 3   0   0.00   8   0   0.00   13   0   0.00
 4   1,117   1,445.78   %129.43   9   0   0.00   14   0   0.00
 5   1,231   2,066.85   %167.9   10   0

007 -234 CONC TOWER - SLABS - MACHINE ROOMS
TYPE BUDGET   ACTUAL   PERCENT   TYPE   BUDGET   ACTUAL   PERCENT   TYPE   BUDGET   ACTUAL PERCENT
 1   0   0.00   6   0   0.00   11   0   0.00
 2   155   142.0   91.61   7   0   0.00   12   0   0.00
 3   0   0.00   8   0   0.00   13   0   0.00
 4   611   695.73   %113.86   9   0   0.00   14   0   0.00
 5   685   643.01   93.87   10   0

007 -235 CONC TOWER -4IN. TOP 2ND FLR
TYPE BUDGET   ACTUAL   PERCENT   TYPE   BUDGET   ACTUAL   PERCENT   TYPE   BUDGET   ACTUAL PERCENT
 1   0   0.00   6   0   0.00   11   0   0.00
 2   26   0.0   7   0   0.00   12   0   0.00
 3   0   0.00   8   0   0.00   13   0   0.00
 4   94   0.00   9   0   0.00   14   0   0.00
 5   137   0.00   10   0

007 -236 CONC TOWER - 4" TOPPING 3RD FL
TYPE BUDGET   ACTUAL   PERCENT   TYPE   BUDGET   ACTUAL   PERCENT   TYPE   BUDGET   ACTUAL PERCENT
 1   0   0.00   6   0   0.00   11   0   0.00
 2   19   36.5   %192.10   7   0   0.00   12   0   0.00
 3   0   0.00   8   0   0.00   13   0   0.00
 4   94   269.42   %286.61   9   0   0.00   14   0   0.00
 5   137   671.21   %489.93   10   0

```
RUN DATE: 25-JUL-88        COST CODE MASTER FILE PRINT-OUT  PAGE 31
JOB NAME :

--------------COST CODE--------------
  NO     DESCRIPTION

007 -237  CONC TOWER - 4" TOPPING 4TH FL
TYPE  BUDGET    ACTUAL    PERCENT    TYPE   BUDGET    ACTUAL   PERCENT    TYPE   BUDGET   ACTUAL  PERCENT
  1      0       0.00                  6      0       38.46               11       0       0.00
  2     18     107.0     %594.44       7      0        0.00               12       0       0.00
  3      0       0.00                  8      0        0.00               13       0       0.00
  4     94     874.74    %930.57       9      0        0.00               14       0       0.00
  5    137   1,044.70    %762.55      10      0        0.00              * * * * *

007 -238  CONC TOWER - 4" TOPPING P/H UPPER LEVEL
TYPE  BUDGET    ACTUAL    PERCENT    TYPE   BUDGET    ACTUAL   PERCENT    TYPE   BUDGET   ACTUAL  PERCENT
  1      0       0.00                  6      0        0.00               11       0       0.00
  2     89     114.0     %128.09       7      0        0.00               12       0       0.00
  3      0       0.00                  8      0        0.00               13       0       0.00
  4    141   2,429.74   %1723.2        9      0        0.00               14       0       0.00
  5    547   2,181.10    %398.73      10      0        0.00              * * * * *

007 -239  CONC TOWER - 5" L/W TOPPING 4TH FLOOR
TYPE  BUDGET    ACTUAL    PERCENT    TYPE   BUDGET    ACTUAL   PERCENT    TYPE   BUDGET   ACTUAL  PERCENT
  1      0       0.00                  6      0        0.00               11       0       0.00
  2     93      58.5      62.90        7      0        0.00               12       0       0.00
  3      0       0.00                  8      0        0.00               13       0       0.00
  4    141     363.30    %257.66       9      0        0.00               14       0       0.00
  5    547     292.00     53.38       10      0        0.00              * * * * *

007 -240  HOIST PAD
TYPE  BUDGET    ACTUAL    PERCENT    TYPE   BUDGET    ACTUAL   PERCENT    TYPE   BUDGET   ACTUAL  PERCENT
  1      0       0.00                  6      0      962.77               11       0       0.00
  2      0      26.0                   7      0        0.00               12       0       0.00
  3      0       0.00                  8      0        0.00               13       0       0.00
  4      0      68.19                  9      0        0.00               14       0       0.00
  5      0       0.00                 10      0        0.00              * * * * *

007 -310  CONC TOWER - BEAMS - HIGH ROOF
TYPE  BUDGET    ACTUAL    PERCENT    TYPE   BUDGET    ACTUAL   PERCENT    TYPE   BUDGET   ACTUAL  PERCENT
  1      0       0.00                  6      0        0.00               11       0       0.00
  2     12      27.0      %225         7      0        0.00               12       0       0.00
  3      0       0.00                  8      0        0.00               13       0       0.00
  4    141     155.44    %110.24       9      0        0.00               14       0       0.00
  5      0       0.00                 10      0        0.00              * * * * *
```

```
RUN DATE: 25-JUL-88        COST CODE MASTER FILE PRINT-OUT  PAGE 32

JOB NAME :

------------COST CODE------------
  NO      DESCRIPTION
--------------------------------

007 -311  CONC TOWER - BEAMS - PARAPET
TYPE  BUDGET      ACTUAL     PERCENT    TYPE  BUDGET    ACTUAL   PERCENT    TYPE  BUDGET    ACTUAL   PERCENT
 1       0         0.00                   6     0                 0.00        11      0                0.00
 2       8        30.0       %375         7     0                 0.00        12      0                0.00
 3       0         0.00                   8     0                 0.00        13      0                0.00
 4      94       286.62      %304.91      9     0                 0.00        14      0                0.00
 5       0         0.00                  10     0                 0.00        *****

008 -101  DEWATER - SYSTEM INSTALLATION
TYPE  BUDGET      ACTUAL     PERCENT    TYPE  BUDGET    ACTUAL   PERCENT    TYPE  BUDGET    ACTUAL   PERCENT
 1     500       168.34       33.67       6     0                 0.00        11      0                0.00
 2       0         0.0                    7     0               121.60        12      0                0.00
 3   1,000     1,449.00      %144.9       8     0                 0.00        13      0                0.00
 4     500       320.62       64.12       9     0                 0.00        14      0                0.00
 5       0         0.00                  10     0                 0.00        *****

008 -102  DEWATER - SYSTEM REMOVAL
TYPE  BUDGET      ACTUAL     PERCENT    TYPE  BUDGET    ACTUAL   PERCENT    TYPE  BUDGET    ACTUAL   PERCENT
 1       0         0.00                   6     0                 0.00        11      0                0.00
 2       0         0.00                   7     0               226.88        12      0                0.00
 3     500       108.00       21.60       8     0                 0.00        13      0                0.00
 4     500       104.04       20.81       9     0                 0.00        14      0                0.00
 5       0         0.00                  10     0                 0.00        *****

008 -103  DEWATER - 2" PUMPS
TYPE  BUDGET      ACTUAL     PERCENT    TYPE  BUDGET    ACTUAL   PERCENT    TYPE  BUDGET    ACTUAL   PERCENT
 1       0        15.00                   6     0                 0.00        11      0                0.00
 2       0         0.0                    7     0                 0.00        12      0                0.00
 3       0     1,615.40                   8     0                 0.00        13      0                0.00
 4   1,000       327.25       32.73       9     0                 0.00        14      0                0.00
 5       0         0.00                  10     0                 0.00        *****

008 -104  DEWATER - MAIN PUMP
TYPE  BUDGET      ACTUAL     PERCENT    TYPE  BUDGET    ACTUAL   PERCENT    TYPE  BUDGET    ACTUAL   PERCENT
 1     500       150.00       30.00       6     0                 0.00        11      0                0.00
 2   2,475     2,484.83      %100.39      7  4,200   3,654.64    87.02        12      0                0.00
 3       0         0.00                   8     0                 0.00        13      0                0.00
 4       0         0.00                   9     0                 0.00        14      0                0.00
 5       0         0.00                  10     0                 0.00        *****
```

RUN DATE: 25-JUL-88   COST CODE MASTER FILE PRINT-OUT PAGE 33

JOB NAME :

---------------COST CODE---------------
 NO   DESCRIPTION

009 -101   EARTHWORK - RAFT EXCAVATE
TYPE	BUDGET	ACTUAL	PERCENT	TYPE	BUDGET	ACTUAL	PERCENT
1	0	0.00		6	0	0.00	
2	0	2.0		7	0	0.00	
3	0	7,099.00		8	0	0.00	
4	500	765.90	%153.18	9	0	0.00	
5	0	0.00		10	0	0.00	
				*****			
				11	0	0.00	
				12	0	0.00	
				13	0	0.00	
				14	0	0.00	
				*****			

009 -102   EARTHWORK - RAFT BACKFILL TO GROUND FLOOR
TYPE	BUDGET	ACTUAL	PERCENT	TYPE	BUDGET	ACTUAL	PERCENT
1	0	0.00		6	0	66.12	
2	0	0.0		7	0	0.00	
3	8,800	6,939.00	78.85	8	0	0.00	
4	6,200	2,508.86	41.81	9	0	0.00	
5	0	0.00		10	0	0.00	
				*****			
				11	0	0.00	
				12	0	0.00	
				13	0	0.00	
				14	0	0.00	
				*****			

009 -103   EARTHWORK - T/H EXCAVATION PADS & FOOTINGS
TYPE	BUDGET	ACTUAL	PERCENT	TYPE	BUDGET	ACTUAL	PERCENT
1	0	0.00		6	0	0.00	
2	0	8.0		7	0	0.00	
3	1,543	1,314.00	85.16	8	0	0.00	
4	0	340.84		9	0	0.00	
5	0	0.00		10	0	0.00	
				*****			
				11	0	0.00	
				12	0	0.00	
				13	0	0.00	
				14	0	0.00	
				*****			

009 -104   EARTHWORK - T/H BACKFILL TO GROUND FLOOR
TYPE	BUDGET	ACTUAL	PERCENT	TYPE	BUDGET	ACTUAL	PERCENT
1	0	0.00		6	0	0.00	
2	0	0.0		7	0	0.00	
3	3,843	3,792.50	98.69	8	0	0.00	
4	2,400	2,622.96	%109.29	9	0	0.00	
5	0	0.00		10	0	0.00	
				*****			
				11	0	0.00	
				12	0	0.00	
				13	0	0.00	
				14	0	0.00	
				*****			

009 -105   EARTHWORK - GARAGE EXCAVATE PADS & FOOTINGS
TYPE	BUDGET	ACTUAL	PERCENT	TYPE	BUDGET	ACTUAL	PERCENT
1	0	0.00		6	0	0.00	
2	0	0.00		7	0	0.00	
3	4,704	8,631.50	%183.49	8	0	0.00	
4	0	607.40		9	0	0.00	
5	0	0.00		10	0	0.00	
				*****			
				11	0	0.00	
				12	0	0.00	
				13	0	0.00	
				14	0	0.00	
				*****			

RUN DATE: 25-JUL-88   COST CODE MASTER FILE PRINT-OUT PAGE 34

JOB NAME :

----------COST CODE----------
    NO   DESCRIPTION

0009 -106 EARTHWORK - GARAGE BACKFILL TOG.F. - PADS & FOOTINGS
TYPE   BUDGET   ACTUAL   PERCENT   TYPE   BUDGET   ACTUAL   PERCENT

 1   0   0.00   6   0   0.00
 2   0   0.00   7   0   0.00
 3   20,641   10,897.50   52.80   8   0   0.00
 4   7,400   3,472.90   46.93   9   0   0.00
 5   0   0.00   10   0   72.66

                                       * * * * *   11   0   0.00
                                                       12   0   0.00
                                                       13   0   0.00
                                                       14   0   0.00

0009 -107 EARTHWORK - GARAGE 2ND LEVEL FILL
TYPE   BUDGET   ACTUAL   PERCENT   TYPE   BUDGET   ACTUAL   PERCENT

 1   0   0.00   6   0   0.00
 2   0   0.00   7   0   0.00
 3   6,800   5,296.00   77.88   8   0   0.00
 4   1,700   3,067.52   %180.44   9   0   0.00
 5   0   0.00   10   0   0.00

                                       * * * * *   11   0   0.00
                                                       12   0   0.00
                                                       13   0   0.00
                                                       14   0   0.00

0009 -108 EARTHWORK - POOL DECK - FILL
TYPE   BUDGET   ACTUAL   PERCENT   TYPE   BUDGET   ACTUAL   PERCENT

 1   0   0.00   6   0   0.00
 2   0   0.00   7   0   0.00
 3   12,000   17,184.60   %143.20   8   0   0.00
 4   1,000   13,408.17   %1340.8   9   0   0.00
 5   0   0.00   10   0   0.00

                                       * * * * *   11   0   0.00
                                                       12   0   0.00
                                                       13   0   0.00
                                                       14   0   0.00

0009 -109 EARTHWORK - PLANTS & SITE - FILL
TYPE   BUDGET   ACTUAL   PERCENT   TYPE   BUDGET   ACTUAL   PERCENT

 1   0   2,528.20   6   0   52.55
 2   0   0.0   7   0   0.00
 3   30,000   31,733.60   %105.77   8   0   0.00
 4   2,000   1,944.15   97.21   9   0   0.00
 5   0   0.00   10   0   0.00

                                       * * * * *   11   0   0.00
                                                       12   0   0.00
                                                       13   0   0.00
                                                       14   0   0.00

0009 -110 EARTHWORK - SITE EXCAVATE - FOOTINGS
TYPE   BUDGET   ACTUAL   PERCENT   TYPE   BUDGET   ACTUAL   PERCENT

 1   0   0.00   6   0   0.00
 2   0   0.0   7   0   0.00
 3   1,500   1,964.00   %130.93   8   0   0.00
 4   0   305.66   9   0   0.00
 5   0   0.00   10   0   0.00

                                       * * * * *   11   0   0.00
                                                       12   0   0.00
                                                       13   0   0.00
                                                       14   0   0.00

RUN DATE: 25-JUL-88   COST CODE MASTER FILE PRINT-OUT PAGE 35

JOB NAME :

------------COST CODE------------
NO   DESCRIPTION

009 -111 EARTHWORK - COMPACTIONS
TYPE	BUDGET	ACTUAL	PERCENT	TYPE	BUDGET	ACTUAL	PERCENT	TYPE	BUDGET	ACTUAL	PERCENT
1	0	0.00		6	0	0.00		11	0	0.00	
2	0	0.0		7	0	196.20		12	0	0.00	
3	7,500	6,373.06	84.97	8	0	0.00		13	0	0.00	
4	0	6,013.13		9	0	1,069.91		14	0	0.00	
5	0	0.00		10	0	0.00		* * * * *			

009 -112 EARTHWORK - STABALIZATION
TYPE	BUDGET	ACTUAL	PERCENT	TYPE	BUDGET	ACTUAL	PERCENT	TYPE	BUDGET	ACTUAL	PERCENT
1	6,000	0.00		6	0	0.00		11	0	0.00	
2	0	0.0		7	0	0.00		12	0	0.00	
3	6,000	0.00		8	0	0.00		13	0	0.00	
4	1,000	0.00		9	0	0.00		14	0	0.00	
5	0	0.00		10	0	0.00		* * * * *			

009 -113 3/4" DRAIN ROCK
TYPE	BUDGET	ACTUAL	PERCENT	TYPE	BUDGET	ACTUAL	PERCENT	TYPE	BUDGET	ACTUAL	PERCENT
1	0	2,066.72		6	0	0.00		11	0	0.00	
2	0	0.0		7	0	0.00		12	0	0.00	
3	0	1,602.63		8	0	0.00		13	0	0.00	
4	0	147.72		9	0	0.00		14	0	0.00	
5	0	0.00		10	0	0.00		* * * * *			

009 -114 MIRAFI - DRAINAGE MATERIAL
TYPE	BUDGET	ACTUAL	PERCENT	TYPE	BUDGET	ACTUAL	PERCENT	TYPE	BUDGET	ACTUAL	PERCENT
1	0	4,231.00		6	0	0.00		11	0	0.00	
2	0	0.0		7	0	0.00		12	0	0.00	
3	0	0.00		8	0	0.00		13	0	0.00	
4	0	0.00		9	0	0.00		14	0	0.00	
5	0	0.00		10	0	0.00		* * * * *			

010 -101 GEN COND - FIELD OFF & SHEDS
TYPE	BUDGET	ACTUAL	PERCENT	TYPE	BUDGET	ACTUAL	PERCENT	TYPE	BUDGET	ACTUAL	PERCENT
1	500	142.79	28.56	6	0	127.44		11	0	0.00	
2	0	0.0		7	0	0.00		12	0	0.00	
3	0	0.00		8	0	0.00		13	0	0.00	
4	1,000	0.00		9	3,000	5,134.50	%171.15	14	0	0.00	
5	0	0.00		10	0	0.00		* * * * *			

RUN DATE: 25-JUL-88   COST CODE MASTER FILE PRINT-OUT PAGE 36

JOB NAME :

----------COST CODE----------
  NO   DESCRIPTION

010 -102 GEN COND - OFF EQUIP & SUP
TYPE BUDGET   ACTUAL   PERCENT   TYPE BUDGET   ACTUAL   PERCENT
 1   3,500   8,928.59   %255.10   6   0   0.00   0.00
 2   0   0.0   7   0   0.00   0.00
 3   0   0.00   8   0   153.83   0.00
 4   0   228.69   22.87   9   0   0.00   0.00
 5   0   0.00   10   0
                                    * * * * *   11   0   0.00
                                                   12   0   0.00
                                                   13   0   0.00
                                                   14   0   0.00

010 -103 GEN COND - MOVE TO BUILDING
TYPE BUDGET   ACTUAL   PERCENT   TYPE BUDGET   ACTUAL   PERCENT
 1   1,000   104.75   10.48   6   0   868.68   0.00
 2   0   0.00   7   0   0.00   0.00
 3   0   0.00   8   0   0.00   0.00
 4   1,000   228.69   22.87   9   0   0.00   0.00
 5   0   0.00   10   0
                                    * * * * *   11   0   0.00
                                                   12   0   0.00
                                                   13   0   0.00
                                                   14   0   0.00

010 -104 GEN COND - TELEPHONE
TYPE BUDGET   ACTUAL   PERCENT   TYPE BUDGET   ACTUAL   PERCENT
 1   0   1.05   6   0   0.00   0.00
 2   0   0.00   7   0   0.00   0.00
 3   0   0.00   8   0   0.00   0.00
 4   0   0.00   9   7,000   9,887.25 %141.24
 5   0   0.00   10   0
                                    * * * * *   11   0   0.00
                                                   12   0   0.00
                                                   13   0   0.00
                                                   14   0   0.00

010 -105 GEN COND - WATER & ELEC - UTIL
TYPE BUDGET   ACTUAL   PERCENT   TYPE BUDGET   ACTUAL   PERCENT
 1   0   0.00   6   0   0.00   0.00
 2   0   0.00   7   0   0.00   0.00
 3   0   0.00   8   0   0.00   0.00
 4   0   0.00   9   12,000 19,433.60 %161.94
 5   0   0.00   10   0
                                    * * * * *   11   0   0.00
                                                   12   0   0.00
                                                   13   0   0.00
                                                   14   0   0.00

010 -106 GEN COND - ICE & CUPS
TYPE BUDGET   ACTUAL   PERCENT   TYPE BUDGET   ACTUAL   PERCENT
 1   6,000   5,018.56   83.64   6   0   0.00   0.00
 2   0   0.0   7   0   0.00   0.00
 3   0   8   0   0.00   0.00
 4   20,000   26,269.06 %131.34   9   0   564.71   0.00
 5   0   0.00   10   0
                                    * * * * *   11   0   0.00
                                                   12   0   0.00
                                                   13   0   0.00
                                                   14   0   0.00

RUN DATE: 25-JUL-88   COST CODE MASTER FILE PRINT-OUT PAGE 37

JOB NAME :

----------------COST CODE----------------
NO   DESCRIPTION

010 -107   GEN COND - TEMP TOILETS
TYPE	BUDGET	ACTUAL	PERCENT	TYPE	BUDGET	ACTUAL	PERCENT
1	0	80.04		6	0	0.00	
2	0	0.0		7	0	0.00	
3	0	0.00		8	0	0.00	
4	0	1,336.70		9	5,000	4,518.36	90.37
5	0	0.00		10	0	0.00	
				*11	0	0.00	
				*12	0	0.00	
				*13	0	0.00	
				*14	0	0.00	

010 -108   GEN COND- BARRICADES
TYPE	BUDGET	ACTUAL	PERCENT	TYPE	BUDGET	ACTUAL	PERCENT
1	5,500	13,524.40	%245.89	6	0	22,445.63	
2	0	0.0		7	0	0.00	
3	0	0.00		8	0	104.88	
4	19,000	2,327.28	12.25	9	6,000	5,088.50	84.81
5	0	74.00		10	0	0.00	
				*11	0	0.00	
				*12	0	0.00	
				*13	0	0.00	
				*14	0	0.00	

010 -109   GEN COND - TRASH CHUTES
TYPE	BUDGET	ACTUAL	PERCENT	TYPE	BUDGET	ACTUAL	PERCENT
1	5,000	5,152.57	%103.05	6	0	788.22	
2	0	0.0		7	0	0.00	
3	0	0.00		8	0	0.00	
4	9,000	320.88	3.57	9	0	0.00	
5	0	37.00		10	0	0.00	
				*11	0	0.00	
				*12	0	0.00	
				*13	0	0.00	
				*14	0	0.00	

010 -110   GEN COND - CLNUP DURING CONST
TYPE	BUDGET	ACTUAL	PERCENT	TYPE	BUDGET	ACTUAL	PERCENT
1	0	61.18		6	0	466.86	
2	0	0.0		7	0	0.00	
3	0	48.00		8	0	0.00	
4	20,000	68,877.70	%344.38	9	0	0.00	
5	0	0.00		10	0	0.00	
				*11	0	0.00	
				*12	0	0.00	
				*13	0	0.00	
				*14	0	0.00	

010 -111   GEN COND - TRASH HAULING
TYPE	BUDGET	ACTUAL	PERCENT	TYPE	BUDGET	ACTUAL	PERCENT
1	0	0.00		6	0	0.00	
2	0	0.00		7	0	0.00	
3	0	0.00		8	0	0.00	
4	0	42.04		9	20,000	57,827.00	%289.13
5	0	0.00		10	0	0.00	
				*11	0	0.00	
				*12	0	0.00	
				*13	0	0.00	
				*14	0	0.00	

RUN DATE: 25-JUL-88   COST CODE MASTER FILE PRINT-OUT PAGE 3f

JOB NAME :

```
         ----------COST CODE----------
  NO     DESCRIPTION
```

010 -112 GEN COND - TRASH CONTAINERS

TYPE	BUDGET	ACTUAL	PERCENT	TYPE	BUDGET	ACTUAL	PERCENT	TYPE	BUDGET	ACTUAL	PERCENT
1	0	3.15		6	0	0.00		11	0	0.00	
2	0	0.00		7	0	0.00		12	0	0.00	
3	0	0.00		8	0	0.00		13	0	0.00	
4	0	0.00		9	15,000	9,078.68	60.52	14	0	0.00	
5	0	0.00		10	0	0.00		* *			

010 -113 GEN COND - PUNCH LIST

TYPE	BUDGET	ACTUAL	PERCENT	TYPE	BUDGET	ACTUAL	PERCENT	TYPE	BUDGET	ACTUAL	PERCENT
1	3,000	1,442.09	48.07	6	0	15,262.55		11	0	0.00	
2	0	0.0		7	0	0.00		12	0	0.00	
3	0	0.00		8	0	0.00		13	0	0.00	
4	30,000	483.46	1.61	9	0	0.00		14	0	0.00	
5	0	0.00		10	0	0.00		* *			

010 -115 GEN COND - BROKEN GLASS

TYPE	BUDGET	ACTUAL	PERCENT	TYPE	BUDGET	ACTUAL	PERCENT	TYPE	BUDGET	ACTUAL	PERCENT
1	0	0.00		6	0	0.00		11	0	0.00	
2	0	0.0		7	0	0.00		12	0	0.00	
3	0	0.00		8	0	0.00		13	0	0.00	
4	0	0.00		9	3,000	0.00		14	0	0.00	
5	0	0.00		10	0	0.00		* *			

010 -116 GEN COND - TOOLS & EQUIPMENT

TYPE	BUDGET	ACTUAL	PERCENT	TYPE	BUDGET	ACTUAL	PERCENT	TYPE	BUDGET	ACTUAL	PERCENT
1	10,000	19,829.52	%198.29	6	0	0.00		11	0	0.00	
2	0	0.0		7	0	0.00		12	0	0.00	
3	0	10,874.89		8	0	0.00		13	0	0.00	
4	0	98.48		9	0	642.50		14	0	0.00	
5	0	0.00		10	0	0.00		* *			

010 -117 GEN COND - TOPPING OUT PARTY

TYPE	BUDGET	ACTUAL	PERCENT	TYPE	BUDGET	ACTUAL	PERCENT	TYPE	BUDGET	ACTUAL	PERCENT
1	1,500	195.56	13.04	6	0	0.00		11	0	0.00	
2	0	0.0		7	0	0.00		12	0	0.00	
3	0	0.00		8	0	0.00		13	0	0.00	
4	0	0.00		9	0	0.00		14	0	0.00	
5	0	0.00		10	0	0.00		* *			

RUN DATE: 25-JUL-88   COST CODE MASTER FILE PRINT-OUT PAGE 39

JOB NAME :

------------COST CODE------------
NO   DESCRIPTION

010-118 GEN COND - LAB TO DIST MATS

TYPE	BUDGET	ACTUAL	PERCENT	TYPE	BUDGET	ACTUAL	PERCENT
1	0	0.00		*11	0	1,351.22	0.00
2	0	0.0		*12	0	0.00	0.00
3	0	0.00		*13	0	0.00	0.00
4	20,000	47,296.32	%236.48	*14	0	0.00	0.00
5	0	0.00		*			

010-119 GEN COND - SCAFFOLDING

TYPE	BUDGET	ACTUAL	PERCENT	TYPE	BUDGET	ACTUAL	PERCENT
1	0	0.00		*11	0	0.00	0.00
2	0	0.0		*12	0	0.00	0.00
3	0	0.00		*13	0	1,277.02	0.00
4	0	0.00		*14	0	0.00	0.00
5	0	0.00		*			

010-120 GEN COND - TRUCK EXPENSE

TYPE	BUDGET	ACTUAL	PERCENT	TYPE	BUDGET	ACTUAL	PERCENT
1	5,000	9,382.34	%187.64	*11	0	0.00	0.00
2	0	0.0		*12	0	0.00	0.00
3	0	172.94		*13	0	0.00	0.00
4	0	24.12		*14	0	0.00	0.00
5	0	0.00		*			

010-121 GEN COND - RUBBING & PATCHING

TYPE	BUDGET	ACTUAL	PERCENT	TYPE	BUDGET	ACTUAL	PERCENT
1	1,000	1,098.35	%109.83	*11	0	0.00	0.00
2	0	0.0		*12	0	0.00	0.00
3	0	0.00		*13	0	205.89	0.00
4	12,000	36,379.68	%303.16	*14	0	0.00	0.00
5	0	14,131.02		*			

010-122 GEN COND - SURVEYS & TESTING

TYPE	BUDGET	ACTUAL	PERCENT	TYPE	BUDGET	ACTUAL	PERCENT
1	0	0.00		*11	0	0.00	0.00
2	0	0.00		*12	0	0.00	0.00
3	0	0.00		*13	0	0.00	0.00
4	0	0.00		*14	1,000	225.00	22.50
5	0	0.00		*			

```
RUN DATE: 25-JUL-88         COST CODE MASTER FILE PRINT-OUT  PAGE 40

JOB NAME :

------------COST CODE------------
 NO      DESCRIPTION

010 -123  GEN COND - PROJECT SIGNS
TYPE  BUDGET    ACTUAL   PERCENT  TYPE  BUDGET    ACTUAL   PERCENT  TYPE  BUDGET   ACTUAL  PERCENT
 1      100      86.75    86.75     6      0       259.16              11     0      0.00
 2        0       0.00                7      0       0.00              12     0      0.00
 3        0       0.00                8      0       0.00              13     0      0.00
 4      500     240.25    48.05     9   1,000    1,145.00   X114.5    14     0      0.00
 5        0       0.00               10      0       0.00            *****

010 -124  GEN COND - RENTAL EQUIPMENT
TYPE  BUDGET    ACTUAL   PERCENT  TYPE  BUDGET    ACTUAL   PERCENT  TYPE  BUDGET   ACTUAL  PERCENT
 1        0      65.26                6      0       0.00              11     0      0.00
 2        0       0.00                7      0       0.00              12     0      0.00
 3   13,000   9,745.92    74.97     8      0       0.00              13     0      0.00
 4        0       0.00                9      0    1,089.00              14     0      0.00
 5        0       0.00               10      0       0.00            *****

010 -125  BLOWING OFF DECKS
TYPE  BUDGET    ACTUAL   PERCENT  TYPE  BUDGET    ACTUAL   PERCENT  TYPE  BUDGET   ACTUAL  PERCENT
 1        0       0.00                6      0       0.00              11     0      0.00
 2        0       0.00                7      0       0.00              12     0      0.00
 3        0       0.00                8      0       0.00              13     0      0.00
 4        0     763.66                9      0       0.00              14     0      0.00
 5        0       0.00               10      0       0.00            *****

010 -127  RUNNER
TYPE  BUDGET    ACTUAL   PERCENT  TYPE  BUDGET    ACTUAL   PERCENT  TYPE  BUDGET   ACTUAL  PERCENT
 1        0      69.45                6      0       0.00              11     0      0.00
 2        0       0.00                7      0    1,237.50              12     0      0.00
 3        0       0.00                8      0       0.00              13     0      0.00
 4        0     370.89                9      0       0.00              14     0      0.00
 5        0       0.00               10      0       0.00            *****

010 -131  INCLEMENT WEATHER
TYPE  BUDGET    ACTUAL   PERCENT  TYPE  BUDGET    ACTUAL   PERCENT  TYPE  BUDGET   ACTUAL  PERCENT
 1        0     623.57                6      0    1,210.47              11     0      0.00
 2        0       0.00                7      0       0.00              12     0      0.00
 3        0       0.00                8      0       0.00              13     0      0.00
 4        0   1,071.28                9      0       0.00              14     0      0.00
 5        0      18.50               10      0       0.00            *****
```

RUN DATE: 25-JUL-88   COST CODE MASTER FILE PRINT-OUT PAGE

JOB NAME :

---------COST CODE---------
 NO   DESCRIPTION

010 -132 CHRISTMAS PARTY

TYPE	BUDGET	ACTUAL	PERCENT	TYPE	BUDGET	ACTUAL	PERCENT
1	0	177.44		6	0	0.00	0.00
2	0	0.00		7	0	0.00	0.00
3	0	0.00		8	0	0.00	0.00
4	0	0.00		9	0	0.00	
5	0	0.00		10	0		

*********NEW ACTIVITY*********

TYPE	BUDGET	ACTUAL	PERCENT
11	0	0.00	
12	0	0.00	
13	0	0.00	
14	0	0.00	
*****			

010 -133 MOVE TO WAREHOUSE

TYPE	BUDGET	ACTUAL	PERCENT	TYPE	BUDGET	ACTUAL	PERCENT	TYPE	BUDGET	ACTUAL	PERCENT
1	0	0.00		6	0	0.00		11	0	0.00	
2	0	0.0		7	0	0.00		12	0	0.00	
3	0	0.00		8	0	455.90		13	0	0.00	
4	0	295.52		9	0	0.00		14	0	0.00	
5	0	0.00		10	0			*****			

011 -101 HOISTING - HOUSE ELEVATOR

TYPE	BUDGET	ACTUAL	PERCENT	TYPE	BUDGET	ACTUAL	PERCENT	TYPE	BUDGET	ACTUAL	PERCENT
1	0	50.00		6	0	27,353.55	x120.5	11	0	0.00	
2	0	0.0		7	22,700	0.00		12	0	0.00	
3	0	0.00		8	0	0.00		13	0	0.00	
4	0	0.00		9	0	0.00		14	0	0.00	
5	0	0.00		10	0			*****			

011 -102 HOISTING - MATERIAL HOIST

TYPE	BUDGET	ACTUAL	PERCENT	TYPE	BUDGET	ACTUAL	PERCENT	TYPE	BUDGET	ACTUAL	PERCENT
1	0	32.71		6	0	63,544.85	71.64	11	0	0.00	
2	0	0.0		7	88,700	0.00		12	0	0.00	
3	121,473	55,645.14	45.81	8	0	4,776.50		13	0	143.55	
4	0	64.69		9	0	0.00		14	0	0.00	
5	0	34.70		10	0			*****			

011 -103 HOISTING - CRANE

TYPE	BUDGET	ACTUAL	PERCENT	TYPE	BUDGET	ACTUAL	PERCENT	TYPE	BUDGET	ACTUAL	PERCENT
1	0	0.00		6	0	1,026.76		11	0	0.00	
2	0	0.0		7	0	0.00		12	0	0.00	
3	505,775	529,958.53	x104.78	8	0	0.00		13	0	0.00	
4	0	0.00		9	0	0.00		14	0	0.00	
5	0	0.00		10	0			*****			

RUN DATE: 25-JUL-88   COST CODE MASTER FILE PRINT-OUT PAGE 44

JOB NAME :

```
------------COST CODE------------
NO     DESCRIPTION
```

013 -103 MILLWORK - H/M DOORS
TYPE	BUDGET	PERCENT	ACTUAL	TYPE	BUDGET	ACTUAL	PERCENT	TYPE	BUDGET	ACTUAL	PERCENT
1	0		18.65	6	7,829	21,791.21	%278.34	* 11	0	0.00	
2	0		0.0	7	0	0.00		* 12	0	0.00	
3	0		0.00	* 8	0	0.00		* 13	0	0.00	
4	0		0.00	* 9	0	0.00		* 14	0	0.00	
5	0		0.00	* 10	0	0.00					

013 -104 MILLWORK - BI-FOLDS
TYPE	BUDGET	PERCENT	ACTUAL	TYPE	BUDGET	ACTUAL	PERCENT	TYPE	BUDGET	ACTUAL	PERCENT
1	0		0.00	6	9,612	14,741.04	%153.36	* 11	0	0.00	
2	0		0.0	7	0	0.00		* 12	0	0.00	
3	0		0.00	* 8	0	0.00		* 13	0	0.00	
4	0		0.00	* 9	0	0.00		* 14	0	0.00	
5	0		0.00	* 10	0	0.00					

013 -105 MILLWORK - MEDICINE CABINETS
TYPE	BUDGET	PERCENT	ACTUAL	TYPE	BUDGET	ACTUAL	PERCENT	TYPE	BUDGET	ACTUAL	PERCENT
1	0		41.20	6	2,607	1,643.28	63.03	* 11	0	0.00	
2	0		0.0	7	0	0.00		* 12	0	0.00	
3	0		0.00	* 8	0	0.00		* 13	0	0.00	
4	0		0.00	* 9	0	0.00		* 14	0	0.00	
5	0		0.00	* 10	0	0.00					

013 -106 MILLWORK - DOOR CLOSERS
TYPE	BUDGET	PERCENT	ACTUAL	TYPE	BUDGET	ACTUAL	PERCENT	TYPE	BUDGET	ACTUAL	PERCENT
1	0		0.00	6	6,024	6,867.02	%113.99	* 11	0	0.00	
2	0		0.0	7	0	0.00		* 12	0	0.00	
3	0		0.00	* 8	0	0.00		* 13	0	0.00	
4	0		0.00	* 9	0	0.00		* 14	0	0.00	
5	0		0.00	* 10	0	0.00					

013 -107 MILLWORK - PANIC DEVICES
TYPE	BUDGET	PERCENT	ACTUAL	TYPE	BUDGET	ACTUAL	PERCENT	TYPE	BUDGET	ACTUAL	PERCENT
1	0		0.00	6	2,656	246.45	9.28	* 11	0	0.00	
2	0		0.0	7	0	0.00		* 12	0	0.00	
3	0		0.00	* 8	0	0.00		* 13	0	0.00	
4	0		0.00	* 9	0	0.00		* 14	0	0.00	
5	0		0.00	* 10	0	0.00					

RUN DATE: 25-JUL-88   COST CODE MASTER FILE PRINT-OUT   PAGE 45

JOB NAME :

----------COST CODE----------
NO   DESCRIPTION

013 -108 MILLWORK - INTERIOR DOOR LOCKS
TYPE	BUDGET	ACTUAL	PERCENT	TYPE	BUDGET	ACTUAL	PERCENT
1	0	0.00		6	7,176	6,997.06	97.51
2	0	0.0		7	0	0.00	
3	0	0.00		8	0	0.00	
4	0	0.00		9	0	0.00	
5	0	0.00		10	0		
				* * * * *			
11	0	0.00					
12	0	0.00					
13	0	0.00					
14	0	0.00					
* * * * *							

013 -109 MILLWORK - THRESHOLDS
TYPE	BUDGET	ACTUAL	PERCENT	TYPE	BUDGET	ACTUAL	PERCENT
1	0	0.00		6	406	640.87	%157.85
2	0	0.0		7	0	0.00	
3	0	0.00		8	0	0.00	
4	0	0.00		9	0	0.00	
5	0	0.00		10	0	0.00	
				* * * * *			
11	0	0.00					
12	0	0.00					
13	0	0.00					
14	0	0.00					
* * * * *							

013 -110 MILLWORK - MAIL BOXES
TYPE	BUDGET	ACTUAL	PERCENT	TYPE	BUDGET	ACTUAL	PERCENT
1	0	0.00		6	200	42.09	21.05
2	0	0.0		7	0	0.00	
3	0	0.00		8	0	0.00	
4	0	0.00		9	0	0.00	
5	0	0.00		10	0	0.00	
				* * * * *			
11	0	0.00					
12	0	0.00					
13	0	0.00					
14	0	0.00					
* * * * *							

013 -111 MILLWORK - BUMPER STOPS
TYPE	BUDGET	ACTUAL	PERCENT	TYPE	BUDGET	ACTUAL	PERCENT
1	0	0.00		6	8,645	0.00	
2	0	0.0		7	0	0.00	
3	0	0.00		8	0	0.00	
4	0	0.00		9	0	0.00	
5	0	0.00		10	0	0.00	
				* * * * *			
11	0	0.00					
12	0	0.00					
13	0	0.00					
14	0	0.00					
* * * * *							

013 -112 MILLWORK - CLOSET SHELVES
TYPE	BUDGET	ACTUAL	PERCENT	TYPE	BUDGET	ACTUAL	PERCENT
1	0	0.00		6	4,792	0.00	
2	0	0.00		7	0	0.00	
3	0	0.00		8	0	0.00	
4	0	0.00		9	0	0.00	
5	0	0.00		10	0	0.00	
				* * * * *			
11	0	0.00					
12	0	0.00					
13	0	0.00					
14	0	0.00					
* * * * *							

RUN DATE: 25-JUL-88   COST CODE MASTER FILE PRINT-OUT PAGE 46

JOB NAME :

```
------------COST CODE------------
 NO     DESCRIPTION
```

013 -113 MILLWORK - TRIM SETS

TYPE	BUDGET	ACTUAL	PERCENT	TYPE	BUDGET	ACTUAL	PERCENT	TYPE	BUDGET	ACTUAL	PERCENT
1	0	0.00		6	3,784	723.60	19.12	11	0	0.00	
2	0	0.00		7	0	0.00		12	0	0.00	
3	0	0.00		8	0	0.00		13	0	0.00	
4	0	0.00		9	0	0.00		14	0	0.00	
5	0	0.00		10	0	0.00		* * * * *			

013 -114 S. C. WOOD DOORS

TYPE	BUDGET	ACTUAL	PERCENT	TYPE	BUDGET	ACTUAL	PERCENT	TYPE	BUDGET	ACTUAL	PERCENT
1	0	0.00		6	457	312.47	68.37	11	0	0.00	
2	0	0.0		7	0	0.00		12	0	0.00	
3	0	0.00		8	0	0.00		13	0	0.00	
4	0	0.00		9	0	0.00		14	0	0.00	
5	0	0.00		10	0	0.00		* * * * *			

013 -115 KICK PLATES

TYPE	BUDGET	ACTUAL	PERCENT	TYPE	BUDGET	ACTUAL	PERCENT	TYPE	BUDGET	ACTUAL	PERCENT
1	0	0.00		6	0	0.00		11	0	0.00	
2	0	0.0		7	0	0.00		12	0	0.00	
3	0	0.00		8	0	0.00		13	0	0.00	
4	0	0.00		9	0	0.00		14	0	0.00	
5	0	0		10	0	0.00		* * * * *			

014 -101 ROUGH CARP - TUBS - FRAMING

TYPE	BUDGET	ACTUAL	PERCENT	TYPE	BUDGET	ACTUAL	PERCENT	TYPE	BUDGET	ACTUAL	PERCENT
1	2,700	4,290.97	%158.92	6	4,000	7,300.81	%182.52	11	0	0.00	
2	0	0.00		7	0	0.00		12	0	0.00	
3	0	0.00		8	0	0.00		13	0	0.00	
4	0	0.00		9	0	0.00		14	0	0.00	
5	0	0.00		10	0	0.00		* * * * *			

014 -102 ROUGH CARP - SUN DECKS

TYPE	BUDGET	ACTUAL	PERCENT	TYPE	BUDGET	ACTUAL	PERCENT	TYPE	BUDGET	ACTUAL	PERCENT
1	3,510	3,415.76	97.32	6	7,695	7,091.37	92.16	11	0	0.00	
2	0	0.0		7	0	0.00		12	0	0.00	
3	0	0.00		8	0	0.00		13	0	0.00	
4	0	110.79		9	0	0.00		14	0	132.24	
5	0	0.00		10	0	0.00		* * * * *			

RUN DATE: 25-JUL-88   COST CODE MASTER FILE PRINT-OUT PAGE 47

JOB NAME :

----------COST CODE----------
  NO   DESCRIPTION
014 -103 ROUGH CARP - SHADOW BOXES
TYPE BUDGET   ACTUAL   PERCENT   TYPE BUDGET   ACTUAL   PERCENT
 1   159   147.52   92.78   6   *****   0.00
 2   0   0.00   7   *****   0.00
 3   0   0.00   8   *****   0.00
 4   0   0.00   9   *****   0.00
 5   0   0.00   10   *****   0.00

                                    11   0   0.00
                                    12   0   0.00
                                    13   0   0.00
                                    14   0   0.00

014 -104 ROUGH CARP - PAV & ROOF DECK
TYPE BUDGET   ACTUAL   PERCENT   TYPE BUDGET   ACTUAL   PERCENT
 1   3,269   6,056.08   %185.25   6   ***** 4,602.58 %191.53
 2   0   0.0   7   *****   0.00
 3   0   0.00   8   *****   0.00
 4   0   0.00   9   *****   0.00
 5   0   151.20   10   *****   0.00

                                    11   0   0.00
                                    12   0   0.00
                                    13   0   0.00
                                    14   0   66.12

014 -105 ROUGH CARP-GAZEBO ROOF & DECK
TYPE BUDGET   ACTUAL   PERCENT   TYPE BUDGET   ACTUAL   PERCENT
 1   4,083   826.53   20.24   6   ***** 3,231.15   87.64
 2   0   0.0   7   *****   0.00
 3   0   0.00   8   *****   0.00
 4   0   0.00   9   *****   0.00
 5   0   0.00   10   *****   0.00

                                    11   0   0.00
                                    12   0   0.00
                                    13   0   0.00
                                    14   0   0.00

014 -106 ROUGH CARP - TRELLIS #1
TYPE BUDGET   ACTUAL   PERCENT   TYPE BUDGET   ACTUAL   PERCENT
 1   1,801   3,155.81   %175.22   6   ***** 5,474.70 %117.76
 2   0   0.0   7   *****   0.00
 3   0   0.00   8   *****   0.00
 4   0   0.00   9   *****   0.00
 5   0   0.00   10   *****   0.00

                                    11   0   0.00
                                    12   0   0.00
                                    13   0   0.00
                                    14   0   132.24

014 -107 ROUGH CARP - TRELLIS #2
TYPE BUDGET   ACTUAL   PERCENT   TYPE BUDGET   ACTUAL   PERCENT
 1   2,494   3,395.16   %136.13   6   ***** 3,388.65 %120.80
 2   0   0.0   7   *****   0.00
 3   0   0.00   8   *****   0.00
 4   0   0.00   9   *****   0.00
 5   0   0.00   10   *****   0.00

                                    11   0   0.00
                                    12   0   0.00
                                    13   0   0.00
                                    14   0   0.00

RUN DATE: 25-JUL-88 COST CODE MASTER FILE PRINT-OUT PAGE 49

JOB NAME :

------------------COST CODE------------------
NO DESCRIPTION

014 -113 DECK OVER A.C. DUCTS

TYPE	BUDGET	ACTUAL	PERCENT	TYPE	BUDGET	ACTUAL	PERCENT	TYPE	BUDGET	ACTUAL	PERCENT
1	0	0.00		* 6	0	99.18		* 11	0	0.00	
2	0	0.0		* 7	0	0.00		* 12	0	0.00	
3	0	0.00		* 8	0	0.00		* 13	0	0.00	
4	0	0.00		* 9	0	0.00		* 14	0	0.00	
5	0	0.00		* 10	0	0.00		*			

014 -114 ROOF OVER STAIR #5

TYPE	BUDGET	ACTUAL	PERCENT	TYPE	BUDGET	ACTUAL	PERCENT	TYPE	BUDGET	ACTUAL	PERCENT
1	0	33.58		* 6	0	264.48		* 11	0	0.00	
2	0	0.0		* 7	0	0.00		* 12	0	0.00	
3	0	0.00		* 8	0	0.00		* 13	0	0.00	
4	0	0.00		* 9	0	0.00		* 14	0	0.00	
5	0	0.00		* 10	0	0.00		*			

014 -115 SAFEWAY EXTRA - (SCAFFOLDS)

TYPE	BUDGET	ACTUAL	PERCENT	TYPE	BUDGET	ACTUAL	PERCENT	TYPE	BUDGET	ACTUAL	PERCENT
1	0	88.00		* 6	0	49.59		* 11	0	0.00	
2	0	0.0		* 7	0	0.00		* 12	0	0.00	
3	0	0.00		* 8	0	0.00		* 13	0	0.00	
4	0	24.62		* 9	0	97.34		* 14	0	0.00	
5	0	0.00		* 10	0	0.00		*			

014 -116 CUTTING HOLES IN JAMBS FOR ELECTRICAL STRIKES

TYPE	BUDGET	ACTUAL	PERCENT	TYPE	BUDGET	ACTUAL	PERCENT	TYPE	BUDGET	ACTUAL	PERCENT
1	0	0.00		* 6	0	559.32		* 11	0	0.00	
2	0	0.0		* 7	0	0.00		* 12	0	0.00	
3	0	0.00		* 8	0	0.00		* 13	0	0.00	
4	0	0.00		* 9	0	0.00		* 14	0	0.00	
5	0	0.00		* 10	0	0.00		*			

014 -117 VIEWING PLAZA

TYPE	BUDGET	ACTUAL	PERCENT	TYPE	BUDGET	ACTUAL	PERCENT	TYPE	BUDGET	ACTUAL	PERCENT
1	0	1,573.43		* 6	0	652.40		* 11	0	0.00	
2	0	0.0		* 7	0	0.00		* 12	0	0.00	
3	0	0.00		* 8	0	0.00		* 13	0	0.00	
4	0	168.16		* 9	0	0.00		* 14	0	0.00	
5	0	588.80		* 10	0	0.00		*			

RUN DATE: 25-JUL-88   COST CODE MASTER FILE PRINT-OUT PAGE 50

JOB NAME :

```
-----------COST CODE----------
  NO       DESCRIPTION
```

015-101 PROJECT SUPERINTENDANT

TYPE	BUDGET	ACTUAL	PERCENT	TYPE	BUDGET	ACTUAL	PERCENT
1	0	0.00	0.00	6	0	0.00	
2	0	0.0	0.00	7	0	0.00	
3	0	0.00	0.00	8	0	226,820.37	
4	0	0.00	0.00	9	0	0.00	
5	0	0.00	0.00	10	0	0.00	
				*11	0	0.00	
				*12	0	0.00	
				*13	0	0.00	
				*14	0	0.00	

015-102 SUPER - STRUCT - TOWER & TH

TYPE	BUDGET	ACTUAL	PERCENT	TYPE	BUDGET	ACTUAL	PERCENT
1	0	0.00		6	0	10,959.00	
2	0	0.0		7	0	0.00	
3	0	0.00		8	39,000	20,622.92	52.88
4	0	494.66		9	0	0.00	
5	0	0.00		10	0	0.00	
				*11	0	0.00	
				*12	0	0.00	
				*13	0	0.00	
				*14	0	0.00	

015-103 SUPER - STRUCTURAL - GARAGE

TYPE	BUDGET	ACTUAL	PERCENT	TYPE	BUDGET	ACTUAL	PERCENT
1	0	0.00		6	0	291.42	
2	0	0.00		7	0	0.00	
3	0	0.00		8	18,200	1,514.10	8.32
4	0	0.00		9	0	0.00	
5	0	0.00		10	0	0.00	
				*11	0	0.00	
				*12	0	0.00	
				*13	0	0.00	
				*14	0	0.00	

015-104 SUPER - STRUCT - SITE & PD

TYPE	BUDGET	ACTUAL	PERCENT	TYPE	BUDGET	ACTUAL	PERCENT
1	0	0.00		6	0	96.24	
2	0	0.0		7	0	0.00	
3	0	0.00		8	8,200	21,404.94	%261.03
4	0	0.00		9	0	0.00	
5	0	0.00		10	0	0.00	
				*11	0	0.00	
				*12	0	0.00	
				*13	0	0.00	
				*14	0	0.00	

015-105 SUPER - INSIDE -TOWER & TH

TYPE	BUDGET	ACTUAL	PERCENT	TYPE	BUDGET	ACTUAL	PERCENT
1	0	0.00		6	0	0.00	
2	0	0.00		7	0	0.00	
3	0	0.00		8	52,500	42,661.25	81.26
4	0	0.00		9	0	0.00	
5	0	0.00		10	0	0.00	
				*11	0	0.00	
				*12	0	0.00	
				*13	0	0.00	
				*14	0	6,075.01	

RUN DATE: 25-JUL-88   COST CODE MASTER FILE PRINT-OUT   PAGE 51

JOB NAME :

----------COST CODE----------
NO   DESCRIPTION

015 -106 SUPER - INSIDE-INTERIOR DESIGN

TYPE	BUDGET	ACTUAL	PERCENT	TYPE	BUDGET	ACTUAL	PERCENT	TYPE	BUDGET	ACTUAL	PERCENT
1	0	0.00		6	0	0.00		11	0	0.00	
2	0	0.0		7	0	0.00		12	0	0.00	
3	0	0.00		8	14,000	0.00		13	0	0.00	
4	0	0.00		9	0	0.00		14	0	0.00	
5	0	0.00		10	0	0.00		* * * * *			

015 -107 SUPER - CLERK OF WORKS

TYPE	BUDGET	ACTUAL	PERCENT	TYPE	BUDGET	ACTUAL	PERCENT	TYPE	BUDGET	ACTUAL	PERCENT
1	0	0.00		6	0	0.00		11	0	0.00	
2	0	0.00		7	0	0.00		12	0	0.00	
3	0	0.00		8	32,400	24,600.00	75.93	13	0	0.00	
4	0	0.00		9	0	0.00		14	0	0.00	
5	0	0.00		10	0	0.00		* * * * *			

015 -108 LABOR SUPERVISION

TYPE	BUDGET	ACTUAL	PERCENT	TYPE	BUDGET	ACTUAL	PERCENT	TYPE	BUDGET	ACTUAL	PERCENT
1	0	0.00		6	0	0.00		11	0	0.00	
2	0	0.0		7	0	0.00		12	0	0.00	
3	0	0.00		8	0	30,257.92		13	0	0.00	
4	0	9,057.05		9	0	0.00		14	0	0.00	
5	0	0.00		10	0	0.00		* * * * *			

015 -111 HEAD CLERK OF WORKS

TYPE	BUDGET	ACTUAL	PERCENT	TYPE	BUDGET	ACTUAL	PERCENT	TYPE	BUDGET	ACTUAL	PERCENT
1	0	0.00		6	0	0.00		11	0	0.00	
2	0	0.0		7	0	0.00		12	0	0.00	
3	0	0.00		8	0	18,062.00		13	0	0.00	
4	0	0.00		9	0	0.00		14	0	0.00	
5	0	0.00		10	0	0.00		* * * * *			

016 -101 MISC - BOLLARDS

TYPE	BUDGET	ACTUAL	PERCENT	TYPE	BUDGET	ACTUAL	PERCENT	TYPE	BUDGET	ACTUAL	PERCENT
1	7,920	0.00		6	0	0.00		11	0	0.00	
2	0	0.0		7	0	0.00		12	0	0.00	
3	0	0.00		8	0	0.00		13	0	0.00	
4	1,000	0.00		9	0	0.00		14	0	0.00	
5	0	0.00		10	0	0.00		* * * * *			

RUN DATE: 25-JUL-88   COST CODE MASTER FILE PRINT-OUT PAGE 55
JOB NAME :

----------------COST CODE----------------
 NO   DESCRIPTION

017 -126 CONCRETE - $ VOLUME
TYPE BUDGET   ACTUAL   PERCENT   TYPE BUDGET   ACTUAL   PERCENT   TYPE   BUDGET   ACTUAL   PERCENT
 1   0   1,023,428.55   6   0   0.00   11   0   0.00
 2   0   0.0   7   0   0.00   12   0   0.00
 3   0   0.00   8   0   0.00   13   0   0.00
 4   0   0.00   9   0   0.00   14   0   0.00
 5   0   0.00   10   0   0.00   * * * * *

017 -127 LABOR - $ VOLUME
TYPE BUDGET   ACTUAL   PERCENT   TYPE BUDGET   ACTUAL   PERCENT   TYPE   BUDGET   ACTUAL   PERCENT
 1   0   0.00   6   0   0.00   11   0   0.00
 2   0   0.0   7   0   0.00   12   0   0.00
 3   0   0.00   8   0   0.00   13   0   0.00
 4   0   0.00   9   0   0.00   14   0   0.00
 5   0   0.00   10   0 1,457,696.75   * * * * *

                                                           *‥‥‥‥‥*NEW ACTIVITY*‥‥‥‥‥*
017 -128 PAYROLL TAXES
TYPE BUDGET   ACTUAL   PERCENT   TYPE BUDGET   ACTUAL   PERCENT   TYPE   BUDGET   ACTUAL   PERCENT
 1   0   0.00   6   0   0.00   11   0   0.00
 2   0   0.00   7   0   0.00   12   0   0.00
 3   0   0.00   8   0   0.00   13   0   0.00
 4   0   0.00   9   0   0.00   14   0   0.00
 5   0   0.00   10   0   84,496.57   * * * * *

017 -129 INSURANCE
TYPE BUDGET   ACTUAL   PERCENT   TYPE BUDGET   ACTUAL   PERCENT   TYPE   BUDGET   ACTUAL   PERCENT
 1   0   0.00   6   0   0.00   11   0   0.00
 2   0   0.00   7   0   0.00   12   0   0.00
 3   0   0.00   8   0   0.00   13   0   0.00
 4   0   0.00   9   0   0.00   14   0   0.00
 5   0   0.00   10   0   36,963.95   * * * * *

017 -130 PETTY CASH - PER FM
TYPE BUDGET   ACTUAL   PERCENT   TYPE BUDGET   ACTUAL   PERCENT   TYPE   BUDGET   ACTUAL   PERCENT
 1   0   3,000.00   6   0   0.00   11   0   0.00
 2   0   0.0   7   0   0.00   12   0   0.00
 3   0   0.00   8   0   0.00   13   0   0.00
 4   0   0.00   9   0   0.00   14   0   0.00
 5   0   0.00   10   0   0.00   * * * * *

RUN DATE: 25-JUL-88   COST CODE MASTER FILE PRINT-OUT PAGE 56

JOB NAME :

------COST CODE------
NO   DESCRIPTION

017 -131 CPM SCHEDULE
TYPE	BUDGET	ACTUAL	PERCENT	TYPE	BUDGET	ACTUAL	PERCENT
1	0	6.30		6	0	0.00	
2	0	0.0		7	0	0.00	
3	0	0.00		8	0	0.00	
4	0	0.00		9	0	13,404.67	
5	0	0.00		10	0	0.00	
				*****			
				11	0	0.00	
				12	0	0.00	
				13	0	0.00	
				14	0	0.00	

017 -132 PETTY CASH FUND TO JOBSITE   *************NEW ACTIVITY**************
TYPE	BUDGET	ACTUAL	PERCENT	TYPE	BUDGET	ACTUAL	PERCENT
1	0	300.00-		6	0	0.00	
2	0	0.0		7	0	0.00	
3	0	0.00		8	0	0.00	
4	0	0.00		9	0	0.00	
5	0	0.00		10	0	300.00	
				*****			
				11	0	0.00	
				12	0	0.00	
				13	0	0.00	
				14	0	0.00	

277 COST CODES ON FI

Appendix H
Job Cost Management Printout:
Volume and Dollar Amounts

RUN DATE: 25-JUL-88   MILLMAN CONSTRUCTION CO., INC.   PAGE 1
JOB COST CUMULATIVE TRANSACTION REPORT
JOB NAME :

COST-CODE	DESCRIPTION	TYPE	DATE	REFERENCE	$ AMOUNT	VOL AMOUNT
002-105	GUARD SHACK - EARTHWORK	3	04/08/85	3 SNEAD CANIPE CK 1593	129.00	
	1 ENTRY FOR THIS TYPE			SUB-TOTALS	129.00	0.0
				TOTALS	129.00	0.0
	0 ENTRIES FOR THIS TYPE				0.00	0.0
	1 ENTRY FOR THIS COST CODE				129.00	0.0
002-106	GUARD SHACK - CONC & ACCESS	2	03/14/85	2 22716		10.0
		2	03/26/85	2 15922 - SPLIT		3.0
		2	03/26/85	2 15927		5.0
				SUB-TOTALS	0.00	18.0
	3 ENTRIES FOR THIS TYPE					
	0 ENTRIES FOR THIS TYPE				0.00	0.0
	3 ENTRIES FOR THIS COST CODE				0.00	18.0
002-108	GUARD SHACK - R/H, CARP, MLWRK	1	05/09/85	1 PAN AM LUMBER CK 1657	641.01	
		1	05/09/85	1 FLA LUMBER CK 1672	18.74	
				SUB-TOTALS	659.75	0.0
				TOTALS	659.75	0.0
	2 ENTRIES FOR THIS TYPE					
	2 ENTRIES FOR THIS COST CODE					
003-101	CONC ACC - EXPANSION MATERIAL	1	09/12/83	1 JAR SUPPLIES	31.80	
		1	09/12/83	1 WALTON CONTEXT	277.21	
		1	10/10/83	1 JAR SUPPLIES	54.51	
		1	10/10/83	1 WALTON CONTEXT	116.54	
		1	11/10/83	1 WALTON-CONTEXT	155.40	
		1	12/09/83	1 BC BUILDERS HARDWARE	154.20	
		1	01/10/84	1 JOHN ABELL CORP CK 33	486.36	
		1	02/10/84	1 WALTON CONTEXT - CHEC	85.65	
		1	03/10/84	1 JAR SUPPLIES - CK 296	72.66	
		1	04/10/84	1 WALTON CK 422	1,253.70	
		1	07/10/84	1 SUNSHINE CK 760	808.50	
		1	08/10/84	1 JAR CK 755	49.96	
		1	08/10/84	1 ABELL CK 834	641.81	
		1	10/15/84	1 PAN AM INS CK 1057	72.00	
		1	10/15/84	1 SUNSHINE CK 1049	404.25	
		1	02/08/85	1 M.G. ENT. CK 1469	204.60	
		1	04/08/85	1 SUNSHINE CONT CK 1598	114.45	
		1	05/09/85	1 SUNSHINE CONT CK 1664	73.08	
				SUB-TOTALS	5,056.08	0.0
				TOTALS	5,056.08	0.0
	18 ENTRIES FOR THIS TYPE					
	18 ENTRIES FOR THIS COST CODE					
003-102	CONC ACC - VISQUEEN	1	09/12/83	1 JAR SUPPLIES	145.33	
		1	09/13/83	1 WALTON CONTEXT	151.73	
		1	12/10/83	1 WALTON CONTEXT	99.23	
		1	12/09/83	1 WALTON CONTEXT	25.20	

```
RUN DATE: 25-JUL-88                    JOB COST CUMULATIVE TRANSACTION REPORT                                PAGE   2
JOB NAME :

COST-CODE   DESCRIPTION              TYPE   DATE      REFERENCE                        $ AMOUNT       VOL  AMOUNT

                                       1   01/10/84 1 JAR SUPPLIES CK 3332                 28.88
                                       1   02/10/84 1 BC BLDRS - CHECK #175              122.98
                                       1   02/10/84 1 WALTON CONTEXT - CHEC              291.38
                                       1   04/10/84 1 JAR CK 466                         144.38
                                       1   08/10/84 1 JAR CK 835                         317.64
                                       1   09/10/84 1 JAR CK 917                          28.67
                                       1   11/09/84 1 BC BLDRS CK 1138                    13.33
                                       1   03/10/85 1 B.C. BLDRS CK 1520                 293.47
                                       1   03/10/85 1 M.G. ENT. CK 1535                   78.50
                                       1   04/08/85 1 B.C. HDWE CK 1587                   18.18
                                       1   04/08/85 1 SUNSHINE CONT CK 1598               54.97
                                                     SUB-TOTALS                        1,813.87        0.0
                                                     TOTALS                            1,813.87        0.0
           *****************************************************************************************************
003-103    15 ENTRIES FOR THIS TYPE
           15 ENTRIES FOR THIS COST CODE
           CONC ACC - CONCRETE SEALER
                                       1   09/12/83 1 BC BUILDERS                        644.06
                                       1   11/10/83 1 JOHN ABELL CORP.                    72.19
                                       1   11/22/83 1 P/C CHECK #3362                   618.75
                                       1   02/01/84 1 PETTY CASH-CHECK #1                 59.54
                                       1   02/10/84 1 JOHN ABELL CORP - CHE              144.38
                                       1   05/10/84 1 JOHN ABELL CK 545                  212.63
                                       1   09/10/84 1 JOHN ABELL CK 916                   65.63
                                       1   10/15/84 1 SUNSHINE CK 1049                    77.43
                                       1   10/10/85 1 BC BLDRS CK 4732                    33.19
                                                     SUB-TOTALS                        1,927.80        0.0
                                                     TOTALS                            1,927.80        0.0
           *****************************************************************************************************
003-104    9 ENTRIES FOR THIS TYPE
           9 ENTRIES FOR THIS COST CODE
           CONC ACC - STYROFOAM
                                       1   12/09/83 1 WALTON CONTEXT                      58.80
                                       1   04/08/85 1 BIGHAM INS. CK 1605                745.82
                                                     SUB-TOTALS                          804.62        0.0
                                                     TOTALS                              804.62        0.0
           *****************************************************************************************************
003-105    2 ENTRIES FOR THIS TYPE
           2 ENTRIES FOR THIS COST CODE
           CONC ACC - DOVETAIL SLOTS
                                       1   09/12/83 1 JAR SUPPLIES                        75.28
                                       1   09/12/83 1 WALTON CONTEXT                      46.20
                                       1   01/10/84 1 SUNSHINE CONT. SUP.                273.00
                                       1   03/10/84 1 WALTON-CONTEXT - CK 3               97.65
                                       1   04/10/84 1 WALTON CK 472                      162.75
                                                     SUB-TOTALS                          654.88        0.0
                                                     TOTALS                              654.88        0.0
           *****************************************************************************************************
003-106    5 ENTRIES FOR THIS TYPE
           5 ENTRIES FOR THIS COST CODE
           CONC ACC - BRICK FOR REBAR
                                       1   08/04/83 1 LONESTAR FLA                       284.77
                                       1   07/10/84 1 RINKER CK 766                       25.76
                                       1   10/15/84 1 RINKER CK 1054                      34.91
                                                     SUB-TOTALS                          345.44        0.0
                                                     TOTALS                              345.44        0.0
           *****************************************************************************************************
003-107    3 ENTRIES FOR THIS TYPE
           3 ENTRIES FOR THIS COST CODE
           CONC ACC - ANTI-HYDRO
                                       1   06/08/84 1 SUNSHINE CK 692                  1,117.75
```

RUN DATE: 25-JUL-88   JOB COST CUMULATIVE TRANSACTION REPORT   PAGE 3

JOB NAME :

COST-CODE	DESCRIPTION	TYPE	DATE	REFERENCE	$ AMOUNT	VOL AMOUNT
		1	10/15/84 1	SUNSHINE CK 1049	242.55	
		1	04/08/85 1	SUNSHINE CONT CK 1598	228.11	
	3 ENTRIES FOR THIS TYPE			SUB-TOTALS	1,588.41	0.0
		9	06/08/84 9	ANTI HYDRO CO CK 693	150.00	
		9	07/10/84 9	ANTI HYDRO CK 762	148.50	
	2 ENTRIES FOR THIS TYPE			SUB-TOTALS	298.50	0.0
	0 ENTRIES FOR THIS TYPE				0.00	0.0
	5 ENTRIES FOR THIS COST CODE			TOTALS	1,886.91	0.0
003-108	CONC ACC - 6" WATER STOP					
		1	10/10/83 1	BC BUILDERS CK 679	35.11	
		1	06/08/84 1	BC BLDRS CK 679	366.32	
	2 ENTRIES FOR THIS TYPE			SUB-TOTALS	441.43	0.0
	2 ENTRIES FOR THIS COST CODE			TOTALS	441.43	0.0
003-110	CONC ACC - NON-METALLIC GROUT					
		1	10/10/83 1	WALTON CONTEXT	57.75	
		1	07/10/84 1	J. ABELL CK 754	181.34	
	2 ENTRIES FOR THIS TYPE			SUB-TOTALS	239.09	0.0
	2 ENTRIES FOR THIS COST CODE			TOTALS	239.09	0.0
003-112	CONC ACC - KEY COVE					
		1	09/12/83 1	JAR SUPPLIES	170.81	
		1	09/12/83 1	WALTON CONTEXT	35.70	
		1	10/10/83 1	WALTON CONTEXT	140.70	
		1	12/09/83 1	WALTON CONTEXT - CHEC	58.80	
		1	02/10/84 1	WALTON CONTEXT	64.65	
		1	08/10/84 1	SUNSHINE CK 843	92.92	
		1	04/08/85 1	SUNSHINE CONT CK 1598	100.80	
	7 ENTRIES FOR THIS TYPE			SUB-TOTALS	663.78	0.0
	7 ENTRIES FOR THIS COST CODE			TOTALS	663.78	0.0
003-113	2ND FLOOR FLOATING SLAB					
		1	09/10/84 1	MIRON CK 922	74.09	
		1	10/15/84 1	MIRON CK 1048	111.13	
	2 ENTRIES FOR THIS TYPE			SUB-TOTALS	185.22	0.0
		2	09/21/84 2	57425	5.0	
		2	09/21/84 2	57438	2.0	
		2	10/08/84 2	57717	2.0	
	3 ENTRIES FOR THIS TYPE			SUB-TOTALS	0.00	9.0
	5 ENTRIES FOR THIS COST CODE			TOTALS	185.22	9.0
004-100	CONC GARAGE - COL PADS & FTGS					
		2	09/12/83 2	LONESTAR FLA		64.0
		2	12/01/83 2	TICKET 53262		10.0
		2	12/01/83 2	TICKET 53263		10.0
		2	12/01/83 2	TICKET 53266		10.0
		2	12/01/83 2	TICKET 53267		10.0

RUN DATE: 25-JUL-88　　　　　　　　　　JOB COST CUMULATIVE TRANSACTION REPORT　　　　　　　　　　PAGE　4

JOB NAME :

COST-CODE	DESCRIPTION	TYPE	DATE	REFERENCE	$ AMOUNT	VOL AMOUNT
		2	12/02/83	TICKET 53315		10.0
		2	12/02/83	TICKET 53319		10.0
		2	12/02/83	TICKET 53320		10.0
		2	12/02/83	TICKET 53322		5.0
		2	12/02/83	TICKET 53323		10.0
		2	12/08/83	TICKET 20571		10.0
		2	12/08/83	TICKET 53437		10.0
		2	12/08/83	TICKET 53438		6.0
		2	12/08/83	TICKET 53445		10.0
		2	12/09/83	TICKET 53449		10.0
		2	12/09/83	TICKET 53485		10.0
		2	12/09/83	TICKET 53486		10.0
		2	12/09/83	TICKET 53489		10.0
		2	12/13/83	TICKET 53550		4.0
		2	12/14/83	TICKET 53579		10.0
		2	12/15/83	TICKET 53616		10.0
		2	12/15/83	TICKET 53618		10.0
		2	12/16/83	TICKET 53644		10.0
		2	12/16/83	TICKET 53645		10.0
		2	12/16/83	TICKET 53646		10.0
		2	12/21/83	TICKET 53718		10.0
		2	12/21/83	TICKET 53719		10.0
		2	12/21/83	TICKET 53720		10.0
		2	12/21/83	TICKET 53721		10.0
		2	12/21/83	TICKET 53724		10.0
		2	12/21/83	TICKET 53726		10.0
		2	12/21/83	TICKET 53727		10.0
		2	12/21/83	TICKET 53728		10.0
		2	12/21/83	TICKET 53729		10.0
		2	12/21/83	TICKET 53730		10.0
		2	12/22/83	TICKET 53731		10.0
		2	12/22/83	TICKET 53766		10.0
		2	12/22/83	TICKET 53767		10.0
		2	12/22/83	TICKET 53768		10.0
		2	12/22/83	TICKET 53769		4.0
		2	12/22/83	TICKET 53770		2.0
		2	12/22/83	TICKET 53771		10.0
		2	12/30/83	TICKET 54852		10.0
		2	12/30/83	TICKET 54854		10.0
		2	01/03/84	TICKET 54883		10.0
		2	01/03/84	TICKET 54885		10.0
		2	01/03/84	TICKET 54886		10.0
		2	01/03/84	TICKET 54887		7.0
		2	01/03/84	TICKET 54897 - SPLIT		10.0
		2	01/03/84	TICKET 54898		10.0
		2	01/05/84	TICKET 54916		10.0
		2	01/05/84	TICKET 54965		10.0
		2	01/05/84	TICKET 54966		10.0
		2	01/05/84	TICKET 54967		10.0
		2	01/05/84	TICKET 54968		10.0
		2	01/05/84	TICKET 54969		10.0

RUN DATE: 25-JUL-88   JOB COST CUMULATIVE TRANSACTION REPORT   PAGE   5
JOB NAME :

COST-CODE DESCRIPTION	TYPE	DATE	REFERENCE		$ AMOUNT	VOL AMOUNT
	2	01/06/84	TICKET 550006	2		2.5
	2	01/06/84	TICKET 550008	2		10.0
	2	01/06/84	TICKET 550009	2		10.0
	2	01/06/84	TICKET 550010	2		10.0
	2	01/06/84	TICKET 550011	2		10.0
	2	01/06/84	TICKET 550012	2		10.0
	2	01/06/84	TICKET 550013	2		5.0
	2	01/10/84	TICKET 550060	2		10.0
	2	01/10/84	TICKET 550062	2		10.0
	2	01/10/84	TICKET 550063	2		10.0
	2	01/10/84	TICKET 550064	2		10.0
	2	01/10/84	TICKET 550065	2		10.0
	2	01/10/84	TICKET 550066	2		10.0
	2	01/10/84	TICKET 550067	2		10.0
	2	01/10/84	TICKET 550068	2	- SPLIT	1.0
	2	01/12/84	TICKET 55128	2		10.0
	2	01/12/84	TICKET 551129	2		10.0
	2	01/12/84	TICKET 55130	2		10.0
	2	01/12/84	TICKET 55135	2		3.0
	2	01/13/84	TICKET 55161	2	- SPLIT	10.0
	2	01/13/84	TICKET 55166	2		10.0
	2	01/13/84	TICKET 55167	2		10.0
	2	01/13/84	TICKET 55168	2		10.0
	2	01/13/84	TICKET 55173	2		10.0
	2	01/13/84	TICKET 55174	2		10.0
	2	01/19/84	TICKET 55307	2		5.0
	2	01/19/84	TICKET 55308	2		8.0
	2	01/19/84	TICKET 55309	2		7.0
	2	01/19/84	TICKET 55310	2		2.0
	2	03/15/84	TICKET 52676	2		10.0
	2	03/15/84	TICKET 52677	2		10.0
	2	03/19/84	TICKET 52740	2	- SPLIT	10.0
	2	03/21/84	TICKET 52780	2		10.0
	2	06/08/84	TICKET 54187	2		10.0
	2	06/08/84	TICKET 54188	2		10.0
	2	06/08/84	TICKET 54192	2		10.0
	2	06/08/84	TICKET 54193	2		10.0
	2	06/08/84	TICKET 01777	2		10.0
	2	06/11/84	TICKET 54222	2		10.0
	2	06/11/84	TICKET 54223	2		6.0
	2	06/11/84	TICKET 54226	2		10.0
	2	06/12/84	TICKET 54238	2		5.0
	2	06/12/84	TICKET 54239	2		10.0
	2	06/15/84	TICKET 54305	2		10.0
	2	06/15/84	TICKET 54306	2		10.0
	2	06/15/84	TICKET 54307	2		5.0
	2	06/15/84	TICKET 54308	2	- SPLIT	8.0
	2	06/15/84	TICKET 01568	2		10.0
	2	06/18/84	TICKET 54324	2		10.0

```
RUN DATE: 25-JUL-88                          JOB COST CUMULATIVE TRANSACTION REPORT                                    PAGE   6
JOB NAME :

COST-CODE   DESCRIPTION             TYPE   DATE      REFERENCE              $ AMOUNT      VOL AMOUNT

                                     2    06/18/84   TICKET 54325                             10.0
                                     2    06/18/84   TICKET 54326                              5.0
                                     2    06/19/84   TICKET 54332                             10.0
                                     2    06/19/84   TICKET 54333                             10.0
                                     2    06/19/84   TICKET 54334                              4.0
                                     2    06/20/84   TICKET 54352                             10.0
                                     2    06/21/84   TICKET 54371                             10.0
                                     2    06/21/84   TICKET 54373                             10.0
                                     2    06/22/84   TICKET 54398                              8.5
                                     2    07/17/84   TICKET 55983 - SPLIT                      5.0
                                     2    07/17/84   TICKET 55987 - SPLIT                      1.5
                                                     -------------------------------------------------
                                                     SUB-TOTALS                 0.00       1,111.5
                                                     TOTALS                     0.00       1,111.5
            ****************************************************************************************
004-102                              2    12/15/83   TICKET 53592                              5.0
                                     2    12/15/83   TICKET 53613 - SPLIT                      2.0
                                     2    12/15/83   TICKET 53648                              5.0
                                     2    12/19/83   TICKET 53664                              7.0
                                     2    12/21/83   TICKET 53725                              5.0
                                     2    12/27/83   TICKET 53790                              7.0
                                     2    12/28/83   TICKET 53799                              7.0
                                     2    12/30/83   TICKET 54858                              2.0
                                     2    01/05/84   TICKET 54963                             10.0
                                     2    01/10/84   TICKET 55069                              8.0
                                     2    01/16/84   TICKET 55201                              5.5
                                     2    01/24/84   TICKET 55401                             10.0
                                     2    03/22/84   TICKET 50818 - SPLIT                     10.0
                                     2    06/12/84   TICKET 54234                             10.0
                                     2    06/15/84   TICKET 54298                              4.0
                                     2    06/19/84   TICKET 54323                              7.0
                                     2    06/20/84   TICKET 54331                              4.0
                                     2    06/22/84   TICKET 54353                              5.0
                                     2    06/22/84   TICKET 54404                              5.0
                                                     -------------------------------------------------
                                                     SUB-TOTALS                 0.00         118.5
                                                     TOTALS                     0.00         118.5
            ****************************************************************************************
            118 ENTRIES FOR THIS TYPE
            118 ENTRIES FOR THIS COST CODE
004-103     CONC GARAGE - VERTS PAD TO 2
                                     2    02/14/84   TICKET 51941                              5.0
                                     2    02/15/84   TICKET 51976                              4.0
                                     2    02/16/84   TICKET 52014                              8.0
                                     2    02/17/84   TICKET 52067                              9.0
                                     2    02/20/84   TICKET 52117                              8.0
                                     2    02/22/84   TICKET 52149                              6.0
                                     2    02/23/84   TICKET 52185 - SPLIT                      3.0
                                     2    03/08/84   TICKET 52493                              9.0
                                     2    03/30/84   TICKET 50951                              6.5
                                     2    04/18/84   TICKET 51260                              5.0
                                     2    04/20/84   TICKET 51315                             10.0
                                     2    01/08/85   06061 - SPLIT                             5.0

            19 ENTRIES FOR THIS TYPE
            19 ENTRIES FOR THIS COST CODE
            CONC GARAGE - VERTS 2 TO 3
```

RUN DATE: 25-JUL-88   JOB COST CUMULATIVE TRANSACTION REPORT   PAGE 7
JOB NAME :

COST-CODE	DESCRIPTION	TYPE	DATE	REFERENCE	$ AMOUNT	VOL AMOUNT
				**		
004-104	12 ENTRIES FOR THIS TYPE			SUB-TOTALS	0.00	75.5
	12 ENTRIES FOR THIS COST CODE			TOTALS	0.00	75.5
	CONC GARAGE - VERTS 3 TO P/D			**		
		2	03/13/84	2 TICKET 52623		8.5
		2	03/15/84	2 TICKET 52674		10.0
		2	03/16/84	2 TICKET 52700		9.0
		2	04/05/84	2 TICKET 51042		10.0
		2	04/06/84	2 TICKET 51084		6.5
		2	04/09/84	2 TICKET 51108		4.5
		2	05/03/84	2 TICKET 51562		5.0
		2	05/03/84	2 TICKET 51563		5.0
		2	05/04/84	2 TICKET 51581		10.0
		2	05/08/84	2 TICKET 51622		5.0
				---		
	10 ENTRIES FOR THIS TYPE			SUB-TOTALS	0.00	73.5
	10 ENTRIES FOR THIS COST CODE			TOTALS	0.00	73.5
004-201	CONC GARAGE - SLAB 1ST LEVEL			**		
		2	01/17/84	2 TICKET 55249		10.0
		2	01/18/84	2 TICKET 55246		10.0
		2	01/18/84	2 TICKET 55247		10.0
		2	01/18/84	2 TICKET 55248		10.0
		2	01/18/84	2 TICKET 55256		10.0
		2	01/18/84	2 TICKET 55262		10.0
		2	01/18/84	2 TICKET 55264		10.0
		2	01/18/84	2 TICKET 55270		10.0
		2	01/18/84	2 TICKET 55271 - SPLIT		3.0
		2	01/20/84	2 TICKET 55322		10.0
		2	01/20/84	2 TICKET 55329		10.0
		2	01/20/84	2 TICKET 55330		10.0
		2	01/20/84	2 TICKET 55331		10.0
		2	01/20/84	2 TICKET 55332		10.0
		2	01/31/84	2 TICKET 55506		10.0
		2	01/31/84	2 TICKET 55507		10.0
		2	01/31/84	2 TICKET 55508		10.0
		2	01/31/84	2 TICKET 55509		10.0
		2	01/31/84	2 TICKET 55515		10.0
		2	01/31/84	2 TICKET 55521		10.0
		2	01/31/84	2 TICKET 55522		10.0
		2	01/31/84	2 TICKET 55524		10.0
		2	01/31/84	2 TICKET 55529		10.0
		2	02/17/84	2 TICKET 52029		10.0
		2	02/17/84	2 TICKET 52030		10.0
		2	02/17/84	2 TICKET 52031		10.0
		2	02/17/84	2 TICKET 52032		10.0
		2	02/17/84	2 TICKET 52033		10.0
		2	02/17/84	2 TICKET 52043		10.0
		2	02/17/84	2 TICKET 52044		10.0
		2	02/17/84	2 TICKET 52047		10.0
		2	02/17/84	2 TICKET 52055		10.0
		2	02/17/84	2 TICKET 52056 - SPLIT		2.0
		2	04/09/84	2 TICKET 51093		10.0

RUN DATE: 25-JUL-84   JOB COST CUMULATIVE TRANSACTION REPORT   PAGE   8

JOB NAME :

COST-CODE   DESCRIPTION   TYPE   DATE   REFERENCE $ AMOUNT   VOL AMOUNT

                                   2   04/09/84 2 TICKET 51094   10.0
                                   2   04/09/84 2 TICKET 51095   10.0
                                   2   04/09/84 2 TICKET 51096   10.0
                                   2   04/09/84 2 TICKET 51097   10.0
                                   2   04/09/84 2 TICKET 51098   10.0
                                   2   04/09/84 2 TICKET 51100   5.0
                                   2   04/09/84 2 TICKET 51103   7.0
                                   2   07/03/84 2 TICKET 54633   11.0
                                   2   07/06/84 2 TICKET 54670   7.0
                                   2   07/06/84 2 TICKET 54672   1.5
                                   2   07/09/84 2 54701   10.0
                                   2   07/09/84 2 54702   10.0
                                   2   07/09/84 2 54704   10.0
                                   2   07/09/84 2 54706   10.0
                                   2   07/09/84 2 54709   10.0
                                   2   07/09/84 2 54711   10.0
                                   2   07/10/84 2 54729   10.0
                                   2   07/10/84 2 54730   10.0
                                   2   07/10/84 2 54733   10.0
                                   2   07/10/84 2 54735   10.0
                                   2   07/10/84 2 54736   10.0
                                   2   07/10/84 2 54737   10.0
                                   2   07/10/84 2 54738   10.0
                                   2   07/10/84 2 54740   10.0
                                   2   07/10/84 2 54743   10.0
                                   2   07/10/84 2 54744   10.0
                                   2   07/10/84 2 54745   10.0
                                   2   07/10/84 2 54748   10.0
                                   2   07/10/84 2 54749   10.0
                                   2   07/10/84 2 54751 - SPLIT   7.0
                                   2   07/20/84 2 56112 -SPLIT   3.0
                                   2   07/20/84 2 56119   10.0
                                   2   07/20/84 2 56121   10.0
                                   2   07/20/84 2 56124   10.0

                                              SUB-TOTALS   0.00   636.5
                                              TOTALS   0.00   636.5
                                   ****
                                   2   07/17/84 2 55940   10.0
                                   2   07/17/84 2 55941   10.0
                                   2   07/17/84 2 55942   10.0
                                   2   07/17/84 2 55943   10.0
                                   2   07/17/84 2 55944   10.0
                                   2   07/17/84 2 55945   10.0
                                   2   07/17/84 2 55946   10.0
                                   2   07/17/84 2 55947   10.0
                                   2   07/17/84 2 55948   10.0
                                   2   07/17/84 2 55949   10.0
                                   2   07/17/84 2 55951   10.0
                                   2   07/17/84 2 55952   10.0
                                   2   07/17/84 2 55953   10.0

           68 ENTRIES FOR THIS TYPE CODE
           68 ENTRIES FOR THIS COST CODE
004-202   CONC GARAGE - SLAB 2ND LEVEL

RUN DATE: 25-JUL-88   JOB COST CUMULATIVE TRANSACTION REPORT   PAGE 9
JOB NAME :

COST-CODE	DESCRIPTION	TYPE	DATE	REFERENCE		$ AMOUNT	VOL AMOUNT
		2	07/17/84	2	55954		10.0
		2	07/17/84	2	55955		10.0
		2	07/17/84	2	55956		10.0
		2	07/17/84	2	55957		10.0
		2	07/17/84	2	55962		10.0
		2	07/17/84	2	55963		10.0
		2	07/17/84	2	55964		10.0
		2	07/17/84	2	55970		10.0
		2	07/17/84	2	55971		10.0
		2	07/17/84	2	55972		10.0
		2	07/17/84	2	55973		10.0
		2	07/17/84	2	55981		10.0
		2	07/17/84	2	55982		10.0
		2	07/17/84	2	55983 -SPLIT		5.0
		2	07/20/84	2	56096		10.0
		2	07/20/84	2	56097		10.0
		2	07/20/84	2	56098		10.0
		2	07/20/84	2	56099		10.0
		2	07/20/84	2	56100		10.0
		2	07/20/84	2	56102		10.0
		2	07/20/84	2	56103		10.0
		2	07/20/84	2	56104		10.0
		2	07/20/84	2	56105		10.0
		2	07/20/84	2	56108		10.0
		2	07/20/84	2	56110		10.0
		2	07/20/84	2	56111		7.0
		2	07/20/84	2	56112 -SPLIT		10.0
		2	07/20/84	2	02141		10.0
		2	07/20/84	2	02143		10.0
		2	07/20/84	2	02144		10.0
		2	07/20/84	2	02146		10.0
		2	07/20/84	2	02147		10.0
		2	07/25/84	2	56208		10.0
		2	07/26/84	2	56209		10.0
		2	07/26/84	2	56210		10.0
		2	07/26/84	2	56211		10.0
		2	07/26/84	2	56212		10.0
		2	07/26/84	2	56213		10.0
		2	07/26/84	2	56214		10.0
		2	07/26/84	2	56215		10.0
		2	07/26/84	2	56216		10.0
		2	07/26/84	2	56217		10.0
		2	07/26/84	2	56218		10.0
		2	07/26/84	2	56219		10.0
		2	07/26/84	2	56220		10.0
		2	07/26/84	2	56221		10.0
		2	07/26/84	2	56222		10.0
		2	07/26/84	2	56223		10.0
		2	07/26/84	2	56224		10.0

RUN DATE: 25-JUL-86
JOB NAME :

JOB COST CUMULATIVE TRANSACTION REPORT

PAGE 10

COST-CODE	DESCRIPTION	TYPE	DATE	REFERENCE	$ AMOUNT	VOL AMOUNT
		2	07/26/84	56225		10.0
		2	07/27/84	56278		10.5
		2	08/01/84	56320		10.0
		2	08/01/84	56321		17.0
				SUB-TOTALS	0.00	679.5
				TOTALS	0.00	679.5
	68 ENTRIES FOR THIS TYPE					
	68 ENTRIES FOR THIS COST CODE					
004-203	CONC GARAGE - SLAB 2ND LEVEL					
		2	02/08/84	TICKET 55731		10.0
		2	02/08/84	TICKET 55732		10.0
		2	02/08/84	TICKET 55733		10.0
		2	02/08/84	TICKET 55734		10.0
		2	02/08/84	TICKET 55735		10.0
		2	02/08/84	TICKET 55755		10.0
		2	02/08/84	TICKET 55756		10.0
		2	02/08/84	TICKET 55757		10.0
		2	02/08/84	TICKET 55758		10.0
		2	02/08/84	TICKET 55759		10.0
		2	02/08/84	TICKET 55771		10.0
		2	02/08/84	TICKET 55772		10.0
		2	02/08/84	TICKET 55778		10.0
		2	02/08/84	TICKET 55779		10.0
		2	02/08/84	TICKET 55783 - SPLIT		7.0
		2	02/15/84	TICKET 51950		10.0
		2	02/15/84	TICKET 51951		10.0
		2	02/15/84	TICKET 51952		10.0
		2	02/15/84	TICKET 51954		10.0
		2	02/15/84	TICKET 51960		10.0
		2	02/15/84	TICKET 51961		10.0
		2	02/15/84	TICKET 51962		10.0
		2	02/15/84	TICKET 51963		10.0
		2	03/12/84	TICKET 52581		10.0
		2	03/12/84	TICKET 52582		10.0
		2	03/12/84	TICKET 52583		10.0
		2	03/12/84	TICKET 52584		10.0
		2	03/12/84	TICKET 52586		10.0
		2	03/12/84	TICKET 52587		10.0
		2	03/12/84	TICKET 52588		10.0
		2	03/12/84	TICKET 52589		10.0
		2	03/12/84	TICKET 52590		10.0
		2	03/12/84	TICKET 52593 - SPLIT		3.0
		2	04/16/84	TICKET 51213		10.0
		2	04/16/84	TICKET 51215		10.0
		2	04/16/84	TICKET 51216		10.0
		2	04/16/84	TICKET 51217		10.0
		2	04/16/84	TICKET 51218		10.0
		2	04/16/84	TICKET 51219		10.0
		2	04/16/84	TICKET 51220		10.0
		2	04/16/84	TICKET 51221		10.0
		2	04/16/84	TICKET 51222		10.0
		2	04/16/84	TICKET 51223		10.0

RUN DATE: 25-JUL-88   JOB COST CUMULATIVE TRANSACTION REPORT   PAGE 11
JOB NAME :

COST-CODE	DESCRIPTION	TYPE	DATE	REFERENCE	$ AMOUNT	VOL AMOUNT
				SUB-TOTALS	0.00	420.0
				TOTALS	0.00	420.0
004-204		2	03/02/84	TICKET 52306		10.0
		2	03/02/84	TICKET 52307		10.0
		2	03/02/84	TICKET 52308		10.0
		2	03/02/84	TICKET 52309		10.0
		2	03/02/84	TICKET 52310		10.0
		2	03/02/84	TICKET 52311		10.0
		2	03/02/84	TICKET 52314		10.0
		2	03/02/84	TICKET 52315		10.0
		2	03/02/84	TICKET 52316		10.0
		2	03/02/84	TICKET 52317		10.0
		2	03/02/84	TICKET 52319		10.0
		2	03/02/84	TICKET 52320		10.0
		2	03/02/84	TICKET 52321		10.0
		2	03/02/84	TICKET 52323		10.0
		2	03/02/84	TICKET 52324		10.0
		2	03/22/84	TICKET 52795		10.0
		2	03/22/84	TICKET 52796		10.0
		2	03/22/84	TICKET 52797		10.0
		2	03/22/84	TICKET 52798		10.0
		2	03/22/84	TICKET 52799		10.0
		2	03/22/84	TICKET 50808		10.0
		2	03/22/84	TICKET 50813		10.0
		2	03/22/84	TICKET 50815		10.0
		2	03/22/84	TICKET 50816		10.0
		2	03/22/84	TICKET 50818 - SPLIT		5.0
		2	05/01/84	TICKET 51464		10.0
		2	05/01/84	TICKET 51465		10.0
		2	05/01/84	TICKET 51466		10.0
		2	05/01/84	TICKET 51467		10.0
		2	05/01/84	TICKET 51468		10.0
		2	05/01/84	TICKET 51476		10.0
		2	05/01/84	TICKET 51478		10.0
		2	05/01/84	TICKET 51479		10.0
		2	05/01/84	TICKET 51480		10.0
		2	05/01/84	TICKET 51481		10.0
		2	05/01/84	TICKET 51483		10.0
		2	05/01/84	TICKET 51485		10.0
		2	05/01/84	TICKET 51486		10.0
		2	05/01/84	TICKET 51488		10.0
		2	05/01/84	TICKET 51490		10.0
				SUB-TOTALS	0.00	395.0
				TOTALS	0.00	395.0
	43 ENTRIES FOR THIS TYPE					
	43 ENTRIES FOR THIS COST CODE					
	CONC GARAGE - SLAB 3RD LEVEL					
004-205		2	05/10/84	TICKET 51663		10.0
		2	05/10/84	TICKET 51667		10.0
		2	05/10/84	TICKET 51668		10.0
	40 ENTRIES FOR THIS TYPE					
	40 ENTRIES FOR THIS COST CODE					
	CONC GARAGE - SLAB POOL DECK					

RUN DATE: 25-JUL-88   JOB COST CUMULATIVE TRANSACTION REPORT   PAGE 12

JOB NAME :

COST-CODE	DESCRIPTION	TYPE	DATE	REFERENCE			$ AMOUNT	VOL AMOUNT
		2	05/10/84	2	TICKET	51669		10.0
		2	05/10/84	2	TICKET	51670		10.0
		2	05/10/84	2	TICKET	51671		10.0
		2	05/10/84	2	TICKET	51673		10.0
		2	05/10/84	2	TICKET	51674		10.0
		2	05/10/84	2	TICKET	51675		10.0
		2	05/10/84	2	TICKET	51677		10.0
		2	05/11/84	2	TICKET	51688		10.0
		2	05/11/84	2	TICKET	51700 - SPLIT			2.0
		2	05/11/84	2	TICKET	51701		10.0
		2	05/11/84	2	TICKET	51702		10.0
		2	05/11/84	2	TICKET	51703		10.0
		2	05/11/84	2	TICKET	51704		10.0
		2	05/11/84	2	TICKET	51705		10.0
		2	05/11/84	2	TICKET	51706		10.0
		2	05/11/84	2	TICKET	51707		10.0
		2	05/11/84	2	TICKET	51708		10.0
		2	05/11/84	2	TICKET	51709		10.0
		2	05/11/84	2	TICKET	51714		6.0
		2	05/11/84	2	TICKET	51722		10.0
		2	05/11/84	2	TICKET	51724		10.0
		2	05/24/84	2	TICKET	53911		10.0
		2	05/24/84	2	TICKET	53912		10.0
		2	05/24/84	2	TICKET	53913		10.0
		2	05/24/84	2	TICKET	53914		10.0
		2	05/24/84	2	TICKET	53915		10.0
		2	05/24/84	2	TICKET	53919		10.0
		2	05/24/84	2	TICKET	53920		10.0
		2	05/24/84	2	TICKET	53921		10.0
		2	05/24/84	2	TICKET	53922		10.0
		2	05/30/84	2	TICKET	53979		10.0
		2	05/30/84	2	TICKET	53980		10.0
		2	05/30/84	2	TICKET	53981		10.0
		2	05/30/84	2	TICKET	53982		10.0
		2	05/30/84	2	TICKET	53983		10.0
		2	05/30/84	2	TICKET	53984		10.0
		2	05/30/84	2	TICKET	53985		10.0
		2	06/04/84	2	TICKET	54028		10.0
		2	06/04/84	2	TICKET	54029		10.0
		2	06/04/84	2	TICKET	54030		10.0
		2	06/04/84	2	TICKET	54031		10.0
		2	06/04/84	2	TICKET	54032		10.0
		2	06/04/84	2	TICKET	54035 - SPLIT			3.0
		2	06/11/84	2	TICKET	54197		10.0
		2	06/11/84	2	TICKET	54198		10.0
		2	06/11/84	2	TICKET	54199		10.0
		2	06/11/84	2	TICKET	54200		10.0
		2	06/11/84	2	TICKET	54202		10.0
		2	06/11/84	2	TICKET	54204		10.0
		2	06/11/84	2	TICKET	54205		10.0
		2	06/11/84	2	TICKET	54206		10.0

RUN DATE: 25-JUL-88   JOB COST CUMULATIVE TRANSACTION REPORT   PAGE 13
JOB NAME :

COST-CODE	DESCRIPTION	TYPE	DATE	REFERENCE	$ AMOUNT	VOL AMOUNT
		2	06/11/84	2 TICKET 54207		10.0
		2	06/11/84	2 TICKET 54208		10.0
		2	06/11/84	2 TICKET 54209		10.0
		2	06/11/84	2 TICKET 54210		10.0
		2	06/11/84	2 TICKET 54211		10.0
		2	06/11/84	2 TICKET 54212		10.0
		2	06/11/84	2 TICKET 54213		10.0
		2	06/11/84	2 TICKET 54214		10.0
		2	06/11/84	2 TICKET 54219		10.0
		2	06/12/84	2 TICKET 54241		8.0
		2	07/11/84	2 54761		10.0
		2	11/09/84	2 58389 - SPLIT		2.0
				SUB-TOTALS	0.00	631.0
				TOTALS	0.00	631.0
004-206	66 ENTRIES FOR THIS TYPE					
	66 ENTRIES FOR THIS COST CODE					
	CONC GARAGE - SLAB					
		2	11/27/84	2 58450		10.0
		2	11/27/84	2 58753		10.0
		2	11/27/84	2 58756		5.0
				SUB-TOTALS	0.00	25.0
				TOTALS	0.00	25.0
004-207	3 ENTRIES FOR THIS TYPE					
	3 ENTRIES FOR THIS COST CODE					
	CONC GARAGE - TOPPING 2ND LEVL					
		2	08/20/84	2 56688		5.0
		2	08/20/84	2 56689		5.0
		2	08/20/84	2 56690		5.0
		2	08/20/84	2 56692		5.0
		2	08/20/84	2 56694		5.0
		2	08/20/84	2 56695		5.0
		2	08/20/84	2 56696		5.0
		2	08/20/84	2 56697		5.0
		2	08/21/84	2 56732		5.0
		2	08/21/84	2 56733		5.0
		2	08/21/84	2 56734		5.0
		2	08/21/84	2 56735		5.0
		2	08/21/84	2 56737		5.0
		2	08/21/84	2 56738		5.0
		2	08/21/84	2 56739		5.0
		2	08/21/84	2 56740		5.0
		2	08/21/84	2 56741		5.0
		2	08/21/84	2 56742		5.0
		2	08/21/84	2 56744		5.0
		2	08/22/84	2 56759		5.0
		2	08/22/84	2 56760		5.0
		2	08/22/84	2 56761		5.0
		2	08/22/84	2 56762		5.0
		2	08/22/84	2 56764		5.0
		2	08/22/84	2 56765		5.0
		2	08/22/84	2 56766		5.0
		2	08/22/84	2 56768		5.0
		2	08/22/84	2 56770		5.0

RUN DATE: 25-JUL-88
JOB NAME :

JOB COST CUMULATIVE TRANSACTION REPORT

PAGE 14

COST-CODE	DESCRIPTION	TYPE	DATE	REFERENCE	$ AMOUNT	VOL AMOUNT
		2	08/22/84	2 56771		5.0
		2	08/22/84	2 56772 - SPLIT		5.0
		2	08/23/84	2 56797		5.0
		2	08/23/84	2 56798		5.0
		2	08/23/84	2 56801		5.0
		2	08/23/84	2 56803		5.0
		2	08/23/84	2 56807		5.0
		2	08/23/84	2 56810		5.0
		2	08/23/84	2 56813		5.0
		2	08/28/84	2 56906		4.0
		2	08/28/84	2 56938		5.0
		2	08/29/84	2 56940		5.0
		2	08/29/84	2 56941		5.0
		2	08/29/84	2 56942		5.0
		2	08/29/84	2 56944		5.0
		2	08/29/84	2 56948		5.0
		2	08/29/84	2 56949		5.0
		2	08/29/84	2 56950		5.0
		2	08/29/84	2 56952		5.0
		2	08/29/84	2 56953		5.0
		2	08/29/84	2 56954		5.0
		2	08/29/84	2 56955		5.0
		2	08/29/84	2 56956		5.0
		2	09/11/84	2 57180		6.0
		2	09/11/84	2 57182		6.0
		2	10/01/84	2 57583		5.0
				SUB-TOTALS	0.00	271.0
				TOTALS	0.00	271.0
		**				
004-208	CONC GARAGE - TOPPING POOL DK	2	08/31/84	2 57031		10.0
		2	09/07/84	2 57108		10.0
		2	09/07/84	2 57109		10.0
		2	09/07/84	2 57113		10.0
		2	09/07/84	2 57116		10.0
		2	09/07/84	2 57117		10.0
		2	09/07/84	2 57118		10.0
		2	09/07/84	2 57119		10.0
		2	09/07/84	2 57127		10.0
		2	09/07/84	2 57136		10.0
		2	09/10/84	2 57151		10.0
		2	09/10/84	2 57152		10.0
		2	09/10/84	2 57153		10.0
		2	09/10/84	2 57156		10.0
		2	09/10/84	2 57157		10.0
		2	09/10/84	2 57158		10.0
		2	09/12/84	2 57196		10.0
		2	09/12/84	2 57197		10.0
		2	09/12/84	2 57198		10.0
		2	09/12/84	2 57199		10.0
		2	09/12/84	2 57202		10.0

54 ENTRIES FOR THIS TYPE
54 ENTRIES FOR THIS COST CODE

RUN DATE: 25-JUL-88   JOB COST CUMULATIVE TRANSACTION REPORT   PAGE 15

JOB NAME :

COST-CODE	DESCRIPTION	TYPE	DATE	REFERENCE	$ AMOUNT	VOL AMOUNT
		2	09/12/84	2 57207		10.0
		2	09/12/84	2 57211		7.0
		2	09/13/84	2 57226		10.0
		2	09/13/84	2 57230		10.0
		2	09/13/84	2 57232		10.0
		2	09/13/84	2 03335		10.0
		2	09/14/84	2 57263		10.0
		2	09/14/84	2 57268		10.0
		2	09/14/84	2 57282		10.0
		2	09/25/84	2 57500		10.0
		2	09/25/84	2 57503		10.0
		2	09/25/84	2 57600		10.0
		2	10/02/84	2 57603		10.0
		2	10/02/84	2 57609		10.0
		2	10/03/84	2 57623		10.0
		2	10/03/84	2 57625		10.0
		2	10/05/84	2 57674		10.0
		2	10/05/84	2 57683		10.0
		2	10/05/84	2 03786		3.0
		2	10/09/84	2 57725		10.0
		2	10/09/84	2 57726		10.0
		2	10/09/84	2 57728		6.0
		2	10/18/84	2 57981		7.0
		2	10/18/84	2 04017		5.0
		2	02/05/85	2 6474		3.5
				SUB-TOTALS	0.00	431.5
				TOTALS	0.00	431.5

46 ENTRIES FOR THIS TYPE
46 ENTRIES FOR THIS COST CODE

| 004-209 | CONC GARAGE - SLAB POOL & WHPL |

		2	05/18/84	2 TICKET 53811		10.0
		2	05/18/84	2 TICKET 53812		10.0
		2	05/18/84	2 TICKET 53813		10.0
		2	05/18/84	2 TICKET 53814		10.0
		2	05/18/84	2 TICKET 53815		10.0
		2	05/18/84	2 TICKET 53816		10.0
		2	05/18/84	2 TICKET 53826		10.0
		2	05/18/84	2 TICKET 53827		10.0
		2	05/18/84	2 TICKET 53828		10.0
		2	05/18/84	2 TICKET 53829		10.0
		2	05/18/84	2 TICKET 53832		10.0
		2	05/18/84	2 TICKET 53833		10.0
		2	05/18/84	2 TICKET 53835		10.0
		2	09/18/84	2 57344 -SPLIT		8.0
				SUB-TOTALS	0.00	138.0
				TOTALS	0.00	138.0

14 ENTRIES FOR THIS TYPE
14 ENTRIES FOR THIS COST CODE

| 004-401 | CONC GARAGE - WALLS 1ST LEVEL |

		2	12/28/83	2 TICKET 54430		7.5
		2	01/03/84	2 TICKET 54898 - SPLIT		3.0
		2	01/03/84	2 TICKET 54901 - SPLIT		6.5
		2	01/04/84	2 TICKET 54936		10.0

RUN DATE: 25-JUL-88   JOB COST CUMULATIVE TRANSACTION REPORT   PAGE 16
JOB NAME :

COST-CODE	DESCRIPTION	TYPE	DATE	REFERENCE	$ AMOUNT	VOL AMOUNT
		2	01/04/84	2 TICKET 54938		4.0
		2	01/05/84	2 TICKET 54972		7.0
		2	01/06/84	2 TICKET 55019 - SPLIT		6.0
		2	01/06/84	2 TICKET 55023		10.0
		2	01/19/84	2 TICKET 55304		10.0
		2	01/24/84	2 TICKET 55397		8.0
		2	03/28/84	2 TICKET 50910		8.0
		2	03/28/84	2 TICKET 50919		6.5
		2	03/29/84	2 TICKET 50927		5.0
		2	03/29/84	2 TICKET 50937		6.5
		2	03/29/84	2 TICKET 50938		11.0
		2	03/29/84	2 TICKET 50939		8.0
		2	04/03/84	2 TICKET 50998		11.0
		2	04/03/84	2 TICKET 50999		11.0
		2	04/03/84	2 TICKET 51000		10.5
		2	06/22/84	2 TICKET 54402		10.0
		2	06/22/84	2 TICKET 54403		10.0
		2	06/26/84	2 TICKET 54446		10.0
		2	06/26/84	2 TICKET 54447		2.0
		2	06/26/84	2 TICKET 54450		10.0
		2	06/27/84	2 TICKET 54472		10.0
		2	06/27/84	2 TICKET 54473		5.0
		2	07/11/84	2 54763 - SPLIT		10.0
		2	07/18/84	2 56004		10.0
		2	07/19/84	2 56087		10.0
		2	07/19/84	2 56089		4.5
				SUB-TOTALS	0.00	241.0
				TOTALS	0.00	241.0

004-402 30 ENTRIES FOR THIS TYPE
30 ENTRIES FOR THIS COST CODE
CONC GARAGE - WALLS 2ND LEVEL

		2	07/11/84	2 54763 - SPLIT		5.0
		2	07/25/84	2 56180 - SPLIT		2.0
		2	07/25/84	2 56182		10.0
		2	08/02/84	2 56341		10.0
		2	08/02/84	2 56342		10.0
		2	08/02/84	2 56345 - SPLIT		1.0
		2	08/02/84	2 56345 - SPLIT		3.0
		2	11/19/84	2 58608		10.0
		2	11/19/84	2 58625		4.0
		2	11/29/84	2 58788 - SPLIT		2.0
		2	11/29/84	2 58786		10.0
		2	12/07/84	2 59001		3.0
		2	12/12/84	2 59288 - SPLIT		8.0
		2	12/27/84	2 59430 - SPLIT		5.0
		2	01/02/85	2 06026		2.0
		2	01/17/85	2 6206275 - SPLIT		
				SUB-TOTALS	0.00	95.0
				TOTALS	0.00	95.0

004-403 16 ENTRIES FOR THIS TYPE
16 ENTRIES FOR THIS COST CODE
CONC GARAGE - WALLS 3RD LEVEL

| | | 2 | 07/11/84 | 2 54765 - SPLIT | | 5.0 |

227

RUN DATE: 25-JUL-88   JOB COST CUMULATIVE TRANSACTION REPORT   PAGE 17

JOB NAME :

COST-CODE	DESCRIPTION	TYPE	DATE	REFERENCE	$ AMOUNT	VOL AMOUNT
		2	07/25/84	56180 -SPLIT		2.0
		2	07/25/84	56181		10.0
		2	08/14/84	56571 - SPLIT		6.0
				SUB-TOTALS	0.00	23.0
				TOTALS	0.00	23.0
004-404	CONC GARAGE - WALLS POOL DECK	2	07/30/83	56291		8.5
		2	12/14/83	TICKET 53590		3.0
		2	02/17/84	TICKET 52065		3.0
		2	02/23/84	TICKET 52185 - SPLIT		5.0
		2	06/15/84	TICKET 54308 - SPLIT		10.0
		2	06/25/84	TICKET 54433		7.0
		2	06/29/84	TICKET 54434		10.0
		2	06/29/84	TICKET 54515		7.0
		2	07/02/84	TICKET 54542		6.0
		2	07/02/84	TICKET 54611		9.0
		2	07/03/84	TICKET 54614		10.0
		2	07/06/84	TICKET 54635		9.0
		2	07/10/84	TICKET 54682		3.0
		2	07/10/84	54751 - SPLIT		7.0
		2	07/11/84	54765 - SPLIT		5.0
		2	07/11/84	54769		10.0
		2	07/11/84	54772		8.0
		2	07/16/84	54774		7.5
		2	07/17/84	55939		1.5
		2	07/24/84	55967 -SPLIT		7.0
		2	07/25/84	56167		6.0
		2	07/26/84	56180 - SPLIT		10.0
		2	07/27/84	56228		8.5
		2	07/31/84	56274		4.0
		2	08/02/84	56313 - SPLIT		5.0
		2	08/02/84	56345		9.0
		2	08/03/84	56348		10.0
		2	08/09/84	56396		11.0
		2	08/13/84	56482		4.0
		2	08/14/84	56557		10.0
		2	08/31/84	56571 - SPLIT		7.0
		2	09/18/84	57013 -SPLIT		2.0
		2	09/18/84	57344 -SPLIT		10.0
		2	09/18/84	57353		10.0
		2	09/18/84	57357		10.0
		2	09/25/84	57505		10.0
		2	10/05/84	57506		10.0
		2	10/10/84	03783		10.0
		2	10/11/84	57788		10.0
		2	10/11/84	57806		10.0
		2	10/11/84	57807		10.0

4 ENTRIES FOR THIS TYPE
4 ENTRIES FOR THIS COST CODE

RUN DATE: 25-JUL-88   JOB COST CUMULATIVE TRANSACTION REPORT   PAGE 18

JOB NAME :

COST-CODE	DESCRIPTION	TYPE	DATE	REFERENCE	$ AMOUNT	VOL AMOUNT
		2	10/12/84	57842		10.0
		2	10/16/84	58033		10.0
		2	10/17/84	57917		10.0
		2	10/17/84	57974		10.0
		2	10/22/84	58034		10.0
		2	10/22/84	58036		10.0
		2	10/29/84	58166		10.0
		2	11/07/84	58341		5.0
		2	11/07/84	58344		8.0
		2	11/07/84	58389 - SPLIT		10.0
		2	11/09/84	58391		7.0
		2	11/09/84	58416 - SPLIT		9.0
		2	11/09/84	58524 - SPLIT		10.0
		2	11/14/84	58526		7.0
		2	11/14/84	58530		6.0
		2	11/19/84	58601 - SPLIT		1.5
		2	12/06/84	58987		5.0
		2	12/28/84	59454 - SPLIT		3.0
		2	01/08/85	06061 - SPLIT		6.0
		2	01/08/85	6071 - SPLIT		5.0
		2	01/18/85	6206277 - SPLIT		1.0
		2	01/25/85	6341 - SPLIT		10.0
		2	01/29/85	6386 - SPLIT		10.0
		2	01/29/85	6388		4.0
		2	01/30/85	6390		10.0
		2	01/30/85	6392		5.0
		2	01/31/85	6408		10.0
		2	01/31/85	6410		3.0
		2	01/31/85	6426		
				SUB-TOTALS	0.00	556.5
				TOTALS	0.00	556.5

74 ENTRIES FOR THIS TYPE
74 ENTRIES FOR THIS COST CODE
004-405   CONC GARAGE - WALLS POOL & WHP

		2	06/04/84	TICKET 54035		7.0
		2	06/04/84	TICKET 54038		10.0
		2	06/04/84	TICKET 54039		10.0
		2	06/04/84	TICKET 54040		5.0
		2	06/04/84	TICKET 54048		5.0
		2	05/28/84	TICKET 01742		10.0
		2	05/28/84	TICKET 01743		10.0
		2	05/28/84	TICKET 54492		3.0
		2	08/31/84	TICKET 57018 - SPLIT		10.0
		2	08/31/84	TICKET 57024		
				SUB-TOTALS	0.00	80.0
				TOTALS	0.00	80.0

10 ENTRIES FOR THIS TYPE
10 ENTRIES FOR THIS COST CODE
005-101   CONC MISC - TIE COLUMNS

		2	12/05/83	TICKET 53346 - SPLIT		2.0
		2	12/05/83	TICKET 53439		4.0
		2	12/14/83	TICKET 53580		3.0

229

RUN DATE: 25-JUL-88   JOB COST CUMULATIVE TRANSACTION REPORT   PAGE 19

JOB NAME :

COST-CODE	DESCRIPTION	TYPE	DATE	REFERENCE		$ AMOUNT	VOL AMOUNT
		2	12/15/83	TICKET 53612			3.0
		2	12/23/83	TICKET 53776			4.0
		2	12/23/83	TICKET 53786			2.0
		2	12/29/83	TICKET 54833			3.0
		2	12/30/83	TICKET 54867			2.0
		2	01/04/84	TICKET 54934			2.5
		2	01/05/84	TICKET 54962			2.5
		2	01/05/84	TICKET 55015			2.5
		2	01/09/84	TICKET 55047			3.0
		2	01/10/84	TICKET 55057			3.0
		2	01/11/84	TICKET 55088			2.5
		2	01/17/84	TICKET 55221			2.5
		2	01/19/84	TICKET 55301			3.0
		2	01/31/84	TICKET 55534			2.5
		2	02/01/84	TICKET 55581			2.5
		2	02/02/84	TICKET 55605 - SPLIT			3.0
		2	02/03/84	TICKET 55630 - SPLIT			8.0
		2	02/06/84	TICKET 55690			3.0
		2	02/07/84	TICKET 55717			3.0
		2	02/08/84	TICKET 55774			4.0
		2	02/10/84	TICKET 51852			2.0
		2	02/13/84	TICKET 51880			3.0
		2	02/16/84	TICKET 52007			3.0
		2	02/17/84	TICKET 52069			3.0
		2	02/20/84	TICKET 52113			3.0
		2	02/24/84	TICKET 52215 - SPLIT			3.0
		2	02/27/84	TICKET 52228			3.0
		2	03/01/84	TICKET 52305 - SPLIT			6.0
		2	03/12/84	TICKET 52593			7.0
		2	03/14/84	TICKET 52632 - SPLIT			4.0
		2	03/16/84	TICKET 52680			4.0
		2	03/19/84	TICKET 52734			3.0
		2	03/20/84	TICKET 52747			2.0
		2	04/04/84	TICKET 51013 - SPLIT			4.0
		2	04/24/84	TICKET 51331			4.0
		2	04/25/84	TICKET 51352			4.0
		2	04/27/84	TICKET 51372			4.0
		2	05/04/84	TICKET 51439			3.0
		2	05/14/84	TICKET 51568 - SPLIT			3.0
		2	05/16/84	TICKET 51748 - SPLIT			2.0
		2	06/05/84	TICKET 51770			1.0
		2	06/14/84	TICKET 54117			3.0
		2	06/22/84	TICKET 54267			1.0
		2	08/29/84	TIKCET 54385			3.0
		2	10/17/84	TICKET 56965			2.5
				TICKET 57924			
				SUB-TOTALS		0.00	152.0
				TOTALS		0.00	152.0

```
*************************************************************
```

005-102 CONC MISC - CELL BLOCK FILLED

49 ENTRIES FOR THIS TYPE CODE
49 ENTRIES FOR THIS COST CODE

| | | 2 | 01/24/84 | 2 TICKET 55355 | | | 2.5 |

RUN DATE: 25-JUL-88   JOB COST CUMULATIVE TRANSACTION REPORT   PAGE 20
JOB NAME :

COST-CODE	DESCRIPTION	TYPE	DATE	REFERENCE	$ AMOUNT	VOL. AMOUNT
		2	01/24/84	2 TICKET 55371		1.0
		2	02/14/84	2 TICKET 51943 - SPLIT		3.0
		2	02/27/84	2 TICKET 52228 - SPLIT		1.0
		2	05/08/84	2 TICKET 51618 - SPLIT		1.0
		2	05/09/84	2 TICKET 51650		2.0
		2	05/15/84	2 TICKET 51763		3.0
		2	06/06/84	2 TICKET 54117 - SPLIT		4.0
		2	08/13/84	2 56560		3.0
				SUB-TOTALS	0.00	20.5
				TOTALS	0.00	20.5
	**
005-103	9 ENTRIES FOR THIS TYPE					
	9 ENTRIES FOR THIS COST CODE					
	CONC MISC - 6" WALL IN TRANS-					
		2	01/17/85	2 6206275 - SPLIT		2.0
		2	01/18/85	2 6206277 - SPLIT		2.0
		2	01/21/85	2 6283		2.0
				SUB-TOTALS	0.00	6.0
				TOTALS	0.00	6.0
	**
005-106	3 ENTRIES FOR THIS TYPE					
	3 ENTRIES FOR THIS COST CODE					
	CONC MISC - FILL CELL BLOCK IN					
		2	10/22/84	2 58055		5.0
		2	10/29/84	2 58174		10.0
		2	10/30/84	2 58198		5.0
		2	10/30/84	2 58205		5.0
		2	10/31/84	2 58209		5.0
		2	10/31/84	2 58216		5.0
		2	11/01/84	2 58224		5.0
		2	11/01/84	2 58244		4.0
		2	11/01/84	2 58268		5.0
		2	11/02/84	2 58279		3.0
		2	11/05/84	2 58293		2.0
		2	11/05/84	2 58315		4.0
		2	11/29/84	2 58788 - SPLIT		2.0
		2	12/27/84	2 59430 - SPLIT		1.0
		2	01/08/85	2 06071 - SPLIT		2.0
		2	01/18/85	2 6206277 - SPLIT		2.0
				SUB-TOTALS	0.00	68.0
				TOTALS	0.00	68.0
	**
005-107	16 ENTRIES FOR THIS TYPE					
	16 ENTRIES FOR THIS COST CODE					
	CONC MISC - FILL CELL BLOCK AT					
		2	01/30/84	2 TICKET 55485		2.0
		2	05/23/84	2 TICKET 53909 - SPLIT		2.0
				SUB-TOTALS	0.00	4.0
	**
	2 ENTRIES FOR THIS TYPE					
	0 ENTRIES FOR THIS TYPE					
005-108	0 ENTRIES FOR THIS COST CODE					
	LOBBY CHANGE					
		1	08/23/84	1 P/C CK 3818	16.28	0.0
				SUB-TOTALS	16.28	0.0
				TOTALS	168.84	0.0
	**
	1 ENTRY FOR THIS TYPE	3	09/10/84	3 BLANCHARD CK 911		

RUN DATE: 25-JUL-88   JOB COST CUMULATIVE TRANSACTION REPORT   PAGE 21
JOB NAME :

COST-CODE	DESCRIPTION	TYPE	DATE	REFERENCE	$ AMOUNT	VOL AMOUNT
	1 ENTRY FOR THIS TYPE			SUB-TOTALS	168.84	0.0
	0 ENTRIES FOR THIS TYPE			SUB-TOTALS	0.00	0.0
	2 ENTRIES FOR THIS COST CODE			TOTALS	185.12	0.0
006-101	CONC SITE & POOL DECK -					
		2	01/25/85	6341 - SPLIT		5.0
		2	01/25/85	6342		10.0
		2	01/29/85	6386 - SPLIT		9.0
		2	01/31/85	6406		10.0
		2	01/31/85	6407		10.0
		2	02/05/85	6473 - SPLIT		5.0
		2	02/12/85	20110 - SPLIT		2.0
	7 ENTRIES FOR THIS TYPE			SUB-TOTALS	0.00	51.0
	7 ENTRIES FOR THIS COST CODE			TOTALS	0.00	51.0
006-102	CONC SITE & POOL DECK					
		2	10/23/84	58060		10.0
		2	10/26/84	58130		10.0
		2	11/14/84	58524 - SPLIT		1.0
		2	02/14/85	20188 - SPLIT		1.0
	4 ENTRIES FOR THIS TYPE			SUB-TOTALS	0.00	22.0
	4 ENTRIES FOR THIS COST CODE			TOTALS	0.00	22.0
006-103	CONC SITE & POOL DECK					
		2	02/05/85	6473 - SPLIT		5.0
		2	02/12/85	20110 - SPLIT		8.0
		2	02/14/85	20188 - SPLIT		2.0
		2	02/14/85	20172		10.0
		2	02/18/85	1057		10.0
		2	02/18/85	1942		10.0
		2	02/21/85	20273		5.0
		2	03/06/85	22521 - SPLIT		2.0
	8 ENTRIES FOR THIS TYPE			SUB-TOTALS	0.00	52.0
	8 ENTRIES FOR THIS COST CODE			TOTALS	0.00	52.0
006-104	CONC SITE & POOL DECK					
		2	07/24/84	56166 -SPLIT		2.0
		2	08/20/84	56721		3.0
		2	01/31/85	6418		2.0
		2	02/21/85	20287		10.0
		2	02/21/85	20287		2.0
		2	02/21/85	20289		10.0
		2	04/18/85	20365 - SPLIT		5.0
		2	04/18/85	19418		6.0
		2	04/30/85	03654		3.0
		2	04/30/85	03650		3.0
		2	05/15/85	14539		8.0
		2	05/20/85	14609		6.0
		2	05/20/85	14618 - SPLIT		2.0

RUN DATE: 25-JUL-88
JOB NAME :

JOB COST CUMULATIVE TRANSACTION REPORT

PAGE 22

COST-CODE	DESCRIPTION	TYPE	DATE	REFERENCE	$ AMOUNT	VOL AMOUNT
				SUB-TOTALS	0.00	63.0
				TOTALS	0.00	63.0
006-105	13 ENTRIES FOR THIS TYPE	2	10/15/84	57870 2 - SPLIT		1.0
	13 ENTRIES FOR THIS COST CODE	2	10/16/84	57901 2 - SPLIT		1.0
	CONC SITE & POOL DECK	2	10/26/84	58131 2 - SPLIT		2.0
		2	03/26/85	15919 2		10.0
		2	03/26/85	15922 2 - SPLIT		7.0
				SUB-TOTALS	0.00	21.0
				TOTALS	0.00	21.0
006-106	5 ENTRIES FOR THIS TYPE	2	07/24/84	56166 2 - SPLIT		5.0
	5 ENTRIES FOR THIS COST CODE	2	10/04/84	57654 2		4.0
	CONC SITE & POOL DECK	2	10/30/84	58102 2 -SPLIT		2.0
		2	11/19/84	58601 2 - SPLIT		3.0
		2	12/12/84	59114 2		1.0
		2	12/12/84	59288 2 - SPLIT		3.0
		2	02/21/85	20365 2 - SPLIT		5.0
		2	02/28/85	20392 2		5.0
		2	02/28/85	20382 2		10.0
		2	03/04/85	20456 2		9.0
		2	03/08/85	22566 2		6.0
		2	03/18/85	22778 2		2.5
		2	03/18/85	22779 2		5.0
		2	03/18/85	22780 2		5.0
		2	03/18/85	22781 2		5.0
		2	03/18/85	22782 2		5.0
		2	03/18/85	22763 2		5.0
		2	03/18/85	22784 2		5.0
		2	03/18/85	22785 2		5.0
		2	03/18/85	22786 2		5.0
		2	03/18/85	22787 2		5.0
		2	03/18/85	22789 2		5.0
		2	03/18/85	22790 2		4.0
		2	03/18/85	22791 2		1.5
		2	03/28/85	15489 2		5.0
		2	04/04/85	00694 2		10.0
		2	04/04/85	19156 2		10.0
		2	04/26/85	03530 2		5.0
		2	04/26/85	03537 2		5.0
		2	05/09/85	03904 2		10.0
		2	05/09/85	03905 2		4.0
		2	05/10/85	03927 2		7.0
		2	06/03/85	14839 2		
				SUB-TOTALS	0.00	171.0
				TOTALS	0.00	171.0
006-108	33 ENTRIES FOR THIS TYPE	2	05/20/85	14618 2 - SPLIT		1.0
	33 ENTRIES FOR THIS COST CODE					
	CONC SITE & POOL DECK					

RUN DATE: 25-JUL-88   JOB COST CUMULATIVE TRANSACTION REPORT   PAGE 23
JOB NAME :

COST-CODE	DESCRIPTION	TYPE	DATE	REFERENCE	$ AMOUNT	VOL AMOUNT
				SUB-TOTALS	0.00	1.0
				TOTALS	0.00	1.0
006-110	CONC SITE & POOL DECK			************************************		
	1 ENTRY FOR THIS TYPE	2	05/02/85	03713		4.0
	1 ENTRY FOR THIS COST CODE			SUB-TOTALS	0.00	4.0
				TOTALS	0.00	4.0
006-111	CONC SITE & POOL DECK - POOL			************************************		
	1 ENTRY FOR THIS TYPE	2	03/06/85	22521 - SPLIT		2.0
	1 ENTRY FOR THIS COST CODE			SUB-TOTALS	0.00	2.0
				TOTALS	0.00	2.0
067-099	CONC TOWER - MUD SLAB			************************************		
	1 ENTRY FOR THIS TYPE	2	08/04/83	2 LONESTAR FLA		126.0
	1 ENTRY FOR THIS COST CODE			SUB-TOTALS	0.00	126.0
				TOTALS	0.00	126.0
007-100	CONC TOWER - RAFT SLAB			************************************		
		2	08/10/83	2 LONESTAR FLA		1,330.0
		2	09/12/83	2 LONESTAR FLA		2,016.0
	2 ENTRIES FOR THIS TYPE			SUB-TOTALS	0.00	3,346.0
		4	08/02/83	4 PETTY CASH	68.90	
	1 ENTRY FOR THIS TYPE			SUB-TOTALS	68.90	0.0
				TOTALS	68.90	3,346.0
007-101	CONC TOWER - PADS & FOOTINGS			************************************		
	0 ENTRIES FOR THIS TYPE					
	3 ENTRIES FOR THIS COST CODE					
		2	09/12/83	2 LONESTAR FLA		178.5
		2	10/10/83	2 LONESTAR FLA		32.0
		2	10/17/83	2 LONESTAR INV 43820		5.0
	3 ENTRIES FOR THIS TYPE			SUB-TOTALS	0.00	215.5
	3 ENTRIES FOR THIS COST CODE			TOTALS	0.00	215.5
007-102	CONC TOWER - VERTS - MAT TO 2			************************************		
		2	09/12/83	2 LONESTAR FLA		326.0
		2	10/10/83	2 LONESTAR FLA		26.0
	2 ENTRIES FOR THIS TYPE			SUB-TOTALS	0.00	352.0
	2 ENTRIES FOR THIS COST CODE			TOTALS	0.00	352.0
007-103	CONC TOWER - VERTS - 2 TO 3			************************************		
		2	10/10/83	2 LONESTAR FLA 43820		213.5
		2	10/17/83	2 LONESTAR INV 44981		8.0
		2	10/26/83	2 LONESTAR INV #46222		3.0
		2	10/31/83	2 LONESTAR INV #46222		24.5
		2	10/31/83	2 LONESTAR INV #46223		2.0
		2	12/05/83	2 TICKET 53346 - SPLIT		2.0
		2	12/05/83	2 TICKET 53353		3.0
		2	01/09/85	2 5114		6.0
		2	03/21/85	2 22828		3.0

RUN DATE: 25-JUL-88
JOB COST CUMULATIVE TRANSACTION REPORT
PAGE 24

JOB NAME :

COST-CODE	DESCRIPTION	TYPE	DATE	REFERENCE	$ AMOUNT	VOL AMOUNT
	9 ENTRIES FOR THIS TYPE			SUB-TOTALS	0.00	265.0
		3	11/10/83	3 R.L. LAPP	100.00	
	1 ENTRY FOR THIS TYPE			SUB-TOTALS	100.00	0.0
007-104	0 ENTRIES FOR THIS TYPE 10 ENTRIES FOR THIS COST CODE CONC TOWER - VERTS - 3 TO 4			TOTALS	100.00	265.0
		2	10/10/83	2 LONESTAR FLA		50.0
		2	10/17/83	2 LONESTAR INVOICE 43820		71.0
		2	10/17/83	2 LONESTAR INV 43821		20.0
		2	10/26/83	2 LONESTAR INV 44981		47.0
		2	10/31/83	2 LONESTAR INV #46222		3.0
		2	10/31/83	2 LONESTAR INV #46223		8.0
		2	11/05/83	2 LONESTAR INVOICE #49104		25.0
		2	11/10/83	2 LONESTAR INV 47484		3.0
		2	11/12/83	2 LONESTAR INVOICE #50039		5.0
	9 ENTRIES FOR THIS TYPE			SUB-TOTALS	0.00	232.0
		3	11/10/83	3 R.L. LAPP	100.00	
	1 ENTRY FOR THIS TYPE			SUB-TOTALS	100.00	0.0
007-105	0 ENTRIES FOR THIS TYPE 10 ENTRIES FOR THIS COST CODE CONC TOWER - VERTS - 4 TO 5			TOTALS	100.00	232.0
		2	10/31/83	2 LONESTAR INV #46222		44.5
		2	11/05/83	2 LONESTAR INVOICE #49104		10.0
		2	11/10/83	2 LONESTAR INV 47484		70.0
		2	11/10/83	2 LONESTAR INV 47485		8.5
		2	12/05/83	2 TICKET 53344		5.0
		2	12/05/83	2 TICKET 53346 - SPLIT		1.0
	6 ENTRIES FOR THIS TYPE			SUB-TOTALS	0.00	139.0
007-106	6 ENTRIES FOR THIS COST CODE CONC TOWER - VERTS - 5 TO 6			TOTALS	0.00	139.0
		2	11/05/83	2 LONESTAR INVOICE #49104		50.0
		2	11/10/83	2 LONESTAR INV 47484		67.2
		2	11/10/83	2 LONESTAR INV 47485		11.5
	3 ENTRIES FOR THIS TYPE			SUB-TOTALS	0.00	128.5
007-107	3 ENTRIES FOR THIS COST CODE CONC TOWER - VERTS - 6 TO 7			TOTALS	0.00	128.5
		2	11/05/83	2 LONESTAR INVOICE #49104		40.0
		2	11/12/83	2 LONESTAR INVOICE #50039		84.5
	2 ENTRIES FOR THIS TYPE			SUB-TOTALS	0.00	124.5
007-108	2 ENTRIES FOR THIS COST CODE VERTICALS 7 TO 8			TOTALS	0.00	124.5
		2	11/11/83	2 TICKET 6252894		10.0

RUN DATE: 25-JUL-88   JOB COST CUMULATIVE TRANSACTION REPORT   PAGE 25
JOB NAME :

COST-CODE	DESCRIPTION	TYPE	DATE	REFERENCE	AMOUNT	VOL AMOUNT
007-109		2	11/14/83	TICKET 52916		10.0
		2	11/14/83	TICKET 52918		10.0
		2	11/14/83	TICKET 52919		10.0
		2	11/14/83	TICKET 52920		10.0
		2	11/14/83	TICKET 52921 - SPLIT		9.5
		2	11/15/83	TICKET 52947		10.0
		2	11/15/83	TICKET 52952		10.0
		2	11/15/83	TICKET 52954		10.0
		2	11/15/83	TICKET 52955		10.0
		2	11/15/83	TICKET 52956		2.5
		2	11/16/83	TICKET 52966		10.0
		2	11/16/83	TICKET 52967		10.0
		2	11/17/83	TICKET 002775 - SPLIT		1.0
				SUB-TOTALS	0.00	123.0
				TOTALS	0.00	123.0

14 ENTRIES FOR THIS TYPE
14 ENTRIES FOR THIS COST CODE
CONC TOWER - VERTS - 8 TO 9

		2	11/17/83	TICKET 53028		10.0
		2	11/17/83	TICKET 53029		10.0
		2	11/18/83	TICKET 53030		7.0
		2	11/18/83	TICKET 53058		10.0
		2	11/18/83	TICKET 53060		10.0
		2	11/18/83	TICKET 53062		10.0
		2	11/18/83	TICKET 53063		10.0
		2	11/18/83	TICKET 53064		10.0
		2	11/18/83	TICKET 53065		4.0
		2	11/21/83	TICKET 53084		10.0
		2	11/21/83	TICKET 53087		10.0
		2	11/21/83	TICKET 53088		8.0
		2	11/22/83	TICKET 53103		10.0
		2	11/22/83	TICKET 53131		4.0
		2	11/22/83	TICKET 53137		10.0
		2	11/22/83	TICKET 53138		10.0
		2	11/22/83	TICKET 53139 - SPLIT		1.0
				SUB-TOTALS	0.00	144.0
				TOTALS	0.00	144.0

17 ENTRIES FOR THIS TYPE
17 ENTRIES FOR THIS COST CODE
CONC TOWER - VERTS - 9 TO 10

007-110		2	11/22/83	TICKET 53139 - SPLIT		9.0
		2	11/23/83	TICKET 53163		10.0
		2	11/23/83	TICKET 53166		10.0
		2	11/23/83	TICKET 53173		10.0
		2	11/28/83	TICKET 53189 - SPLIT		6.0
		2	11/28/83	TICKET 53191		10.0
		2	11/28/83	TICKET 53193		10.0
		2	11/28/83	TICKET 53194		10.0
		2	11/28/83	TICKET 53195		2.0
		2	11/29/83	TICKET 53221 - SPLIT		7.0
		2	11/29/83	TICKET 53223		10.0
		2	11/29/83	TICKET 53224		7.0
		2	12/01/83	TICKET 53260 - PER DA		10.0

RUN DATE: 25-JUL-88
JOB COST CUMULATIVE TRANSACTION REPORT
PAGE 32

JOB NAME :

COST-CODE	DESCRIPTION	TYPE	DATE	REFERENCE	AMOUNT	VOL AMOUNT
		2	03/01/84	TICKET 52298		10.0
		2	03/01/84	TICKET 52301		10.0
		2	03/02/84	TICKET 52325		10.0
		2	03/02/84	TICKET 52328		4.0
		2	03/05/84	TICKET 52357		10.0
		2	03/05/84	TICKET 52359 - SPLIT		7.0
		2	03/05/84	TICKET 52361		10.0
				SUB-TOTALS	0.00	121.0
		4	01/01/83	4 TEST TRANSACTION	1,234.56	
				SUB-TOTALS	1,234.56	0.0
				SUB-TOTALS	0.00	121.0
				TOTALS	1,234.56	
		**				
		2	03/05/84	TICKET 52366		10.0
		2	03/05/84	TICKET 52367		3.0
		2	03/06/84	TICKET 52404 - SPLIT		8.0
		2	03/06/84	TICKET 52407		10.0
		2	03/06/84	TICKET 52408		10.0
		2	03/07/84	TICKET 52442 - SPLIT		2.0
		2	03/07/84	TICKET 52457		10.0
		2	03/07/84	TICKET 52458		10.0
		2	03/08/84	TICKET 52460		10.0
		2	03/08/84	TICKET 52462 - SPLIT		6.0
		2	03/08/84	TICKET 52491		10.0
		2	03/09/84	TICKET 52499		10.0
		2	03/09/84	TICKET 52531		10.0
		2	03/09/84	TICKET 52533		10.0
		2	03/09/84	TICKET 52538 - SPLIT		1.0
				SUB-TOTALS	0.00	120.0
				TOTALS	0.00	120.0
		**				
		2	03/09/84	TICKET 52535		10.0
		2	03/09/84	TICKET 52538 - SPLIT		2.0
		2	03/12/84	TICKET 52595		10.0
		2	03/12/84	TICKET 52596		10.0
		2	03/12/84	TICKET 52598		10.0
		2	03/13/84	TICKET 52624		10.0
		2	03/13/84	TICKET 52626		10.0
		2	03/13/84	TICKET 52627		10.0
		2	03/14/84	TICKET 52649		10.0
		2	03/14/84	TICKET 52653		10.0
		2	03/14/84	TICKET 52657		3.0
		2	03/16/84	TICKET 52691		10.0
		2	03/16/84	TICKET 52694		5.0
		2	03/16/84	TICKET 52696		
				SUB-TOTALS	0.00	120.0

13 ENTRIES FOR THIS TYPE
1 ENTRY FOR THIS TYPE
0 ENTRIES FOR THIS TYPE
14 ENTRIES FOR THIS COST CODE
007-129 CONC TOWER - VERTS - 28 TO 29

15 ENTRIES FOR THIS TYPE
15 ENTRIES FOR THIS COST CODE
007-130 CONC TOWER - VERTS - 29 TO

14 ENTRIES FOR THIS TYPE

RUN DATE: 25-JUL-88   JOB COST CUMULATIVE TRANSACTION REPORT   PAGE 33
JOB NAME :

COST-CODE	DESCRIPTION	TYPE	DATE	REFERENCE	$ AMOUNT	VOL AMOUNT
				TOTALS	0.00	120.0
				**************		
007-131	14 ENTRIES FOR THIS COST CODE					
	CONC TOWER - VERTICALS - TOWER					
		2	03/19/84	2 TICKET 52734 - SPLIT		4.0
		2	03/19/84	2 TICKET 52741		10.0
		2	03/20/84	2 TICKET 52748		10.0
		2	03/21/84	2 TICKET 52790		10.0
		2	03/22/84	2 TICKET 52791		10.0
		2	03/22/84	2 TICKET 50825		10.0
		2	03/22/84	2 TICKET 50827		10.0
		2	03/23/84	2 TICKET 50859 - SPLIT		6.0
		2	03/26/84	2 TICKET 50867		10.0
		2	03/26/84	2 TICKET 50870		4.0
				SUB-TOTALS	0.00	86.0
				TOTALS	0.00	86.0
				**************		
007-132	10 ENTRIES FOR THIS TYPE					
	10 ENTRIES FOR THIS COST CODE					
	CONC TOWER - VERTS - P/H LOWER					
		2	03/30/84	2 TICKET 50961		10.0
		2	03/30/84	2 TICKET 50965		10.0
		2	04/02/84	2 TICKET 50988 - SPLIT		9.0
		2	04/03/84	2 TICKET 51001		10.0
		2	04/03/84	2 TICKET 51002		10.0
		2	04/03/84	2 TICKET 51003		10.0
		2	04/04/84	2 TICKET 51017		10.0
		2	04/05/84	2 TICKET 51045		10.0
		2	04/05/84	2 TICKET 51047		8.0
		2	04/10/84	2 TICKET 51139 - SPLIT		4.0
				SUB-TOTALS	0.00	91.0
				TOTALS	0.00	91.0
				**************		
007-133	10 ENTRIES FOR THIS TYPE					
	10 ENTRIES FOR THIS COST CODE					
	CONC TOWER - VERTS - P/H UPPER					
		2	04/13/84	2 TICKET 51208		10.0
		2	04/13/84	2 TICKET 51209		10.0
		2	04/16/84	2 TICKET 51224		10.0
		2	04/16/84	2 TICKET 51226		8.0
		2	04/16/84	2 TICKET 51232		10.0
		2	04/17/84	2 TICKET 51214		10.0
		2	04/17/84	2 TICKET 51242		5.0
		2	04/18/84	2 TICKET 51247		3.0
		2	04/18/84	2 TICKET 51254		10.0
		2	04/19/84	2 TICKET 51281 - SPLIT		2.0
		2	04/23/84	2 TICKET 51331 - SPLIT		6.0
		2	04/25/84	2 TICKET 51367		5.0
		2	04/25/84	2 TICKET 51359		10.0
		2	04/26/84	2 TICKET 51399 - SPLIT		5.0
		2	04/27/84	2 TICKET 51431		10.0
				SUB-TOTALS	0.00	114.0
				TOTALS	0.00	114.0
				**************		
007-134	15 ENTRIES FOR THIS TYPE					
	15 ENTRIES FOR THIS COST CODE					
	CONC TOWER - VERTS - TO HIGH					
		2	05/04/84	2 TICKET 51575		10.0
		2	05/07/84	2 TICKET 51600		10.0

RUN DATE: 25-JUL-88   JOB COST CUMULATIVE TRANSACTION REPORT   PAGE 34
JOB NAME :

COST-CODE	DESCRIPTION	TYPE	DATE	REFERENCE	$ AMOUNT	VOL AMOUNT
		2	05/07/84	2 TICKET 51601 - SPLIT		6.0
		2	05/08/84	2 TICKET 51618 - SPLIT		4.0
		2	05/14/84	2 TICKET 51757		10.0
		2	05/16/84	2 TICKET 51770 - SPLIT		8.0
	****************** SUB-TOTALS ******************				0.00	48.0
	6 ENTRIES FOR THIS TYPE TOTALS				0.00	48.0
007-135	6 ENTRIES FOR THIS COST CODE
	CONC TOWER - VERTICALS - TO
		2	05/22/84	2 TICKET 53873		10.0
		2	05/23/84	2 TICKET 53909 - SPLIT		4.0
		2	06/01/84	2 TICKET 01327 - SPLIT		3.0
	****************** SUB-TOTALS ******************				0.00	17.0
	3 ENTRIES FOR THIS TYPE TOTALS				0.00	17.0
007-201	3 ENTRIES FOR THIS COST CODE
	CONC TOWER - SLABS - GROUND FL
		2	09/12/83	2 LONESTAR FLA		210.0
		2	10/10/83	2 LONESTAR FLA		142.5
		2	12/01/83	2 TICKET 53249		10.0
		2	12/01/83	2 TICKET 53252		10.0
		2	12/01/83	2 TICKET 53254		10.0
		2	12/01/83	2 TICKET 53258		10.0
	****************** SUB-TOTALS ******************				0.00	392.5
	6 ENTRIES FOR THIS TYPE TOTALS				0.00	392.5
007-202	6 ENTRIES FOR THIS COST CODE
	CONC TOWER - SLABS - 2ND FLOOR
		2	10/10/83	2 LONESTAR INV 43820		564.0
		2	10/17/83	2 LONESTAR INV 43820		130.0
		2	02/23/84	2 TICKET 52185 - SPLIT		4.5
	****************** SUB-TOTALS ******************				0.00	698.5
	3 ENTRIES FOR THIS TYPE TOTALS				0.00	698.5
007-203	3 ENTRIES FOR THIS COST CODE
	CONC TOWER - SLABS - 3RD FLOOR
		2	10/10/83	2 LONESTAR FLA		297.0
		2	10/11/83	2 LONESTAR FLA		10.0
		2	10/17/83	2 LONESTAR INV 43820		186.0
		2	11/05/83	2 LONESTAR INVOICE #49104		130.0
	****************** SUB-TOTALS ******************				0.00	623.0
	4 ENTRIES FOR THIS TYPE TOTALS				0.00	623.0
007-204	4 ENTRIES FOR THIS COST CODE
	CONC TOWER - SLABS - 4TH FLOOR
		2	10/31/83	2 LONESTAR INV #46222		380.0
		2	11/10/83	2 LONESTAR INV 47484		245.0
		2	11/11/83	2 TICKET 6252892		9.0
		2	11/12/83	2 LONESTAR INVOICE #50039		253.0
	****************** SUB-TOTALS ******************				0.00	887.0
	4 ENTRIES FOR THIS TYPE TOTALS				0.00	887.0
007-205	4 ENTRIES FOR THIS COST CODE
	CONC TOWER - SLABS - 5TH FLOOR
		2	11/10/83	2 LONESTAR INV 47484		412.0
	****************** SUB-TOTALS ******************				0.00	412.0
	1 ENTRY FOR THIS TYPE

RUN DATE: 25-JUL-88   JOB COST CUMULATIVE TRANSACTION REPORT   PAGE 35

JOB NAME :

COST-CODE	DESCRIPTION	TYPE	DATE	REFERENCE	$ AMOUNT	VOL AMOUNT
				TOTALS **		412.0
007-206	1 ENTRY FOR THIS COST CODE					
	CONC.TOWER - SLABS - 6TH FLOOR	2	11/05/83	LONESTAR INVOICE #49104		304.0
		2	11/12/83	LONESTAR INVOICE #50039		120.0
				SUB-TOTALS	0.00	424.0
				TOTALS	0.00	424.0
				**		
007-207	2 ENTRIES FOR THIS TYPE					
	2 ENTRIES FOR THIS COST CODE					
	CONC TOWER - SLABS - 7TH FLOOR	2	11/12/83	LONESTAR INVOICE #50039		150.0
		2	11/14/83	TICKET 6252895		10.0
		2	11/14/83	TICKET 6252896		10.0
		2	11/14/83	TICKET 52897		10.0
		2	11/14/83	TICKET 52898		10.0
		2	11/14/83	TICKET 52899		10.0
		2	11/14/83	TICKET 52900		10.0
		2	11/14/83	TICKET 52901		10.0
		2	11/14/83	TICKET 52902		10.0
		2	11/14/83	TICKET 52904		10.0
		2	11/14/83	TICKET 52905		10.0
		2	11/14/83	TICKET 52907		10.0
		2	11/14/83	TICKET 52908		10.0
		2	11/14/83	TICKET 52910		10.0
		2	11/14/83	TICKET 52911		3.5
		2	11/14/83	TICKET 52915		0.5
		2	11/14/83	TICKET 52921 - SPLIT		10.0
		2	11/15/83	TICKET 52922		10.0
		2	11/15/83	TICKET 52923		10.0
		2	11/15/83	TICKET 52924		10.0
		2	11/15/83	TICKET 52925		10.0
		2	11/15/83	TICKET 52926		10.0
		2	11/15/83	TICKET 52927		10.0
		2	11/15/83	TICKET 52930		10.0
		2	11/15/83	TICKET 52932		10.0
		2	11/15/83	TICKET 52935		10.0
		2	11/15/83	TICKET 52936		10.0
		2	11/15/83	TICKET 52937		10.0
		2	11/15/83	TICKET 52938		10.0
		2	11/17/83	TICKET 002775 - SPLIT		5.0
				SUB-TOTALS	0.00	419.0
				TOTALS	0.00	419.0
				**		
007-208	30 ENTRIES FOR THIS TYPE					
	30 ENTRIES FOR THIS COST CODE					
	CONC TOWER - SLABS - 8TH FLOOR	2	11/17/83	TICKET 52968		10.0
		2	11/17/83	TICKET 52969		10.0
		2	11/17/83	TICKET 52970		10.0
		2	11/17/83	TICKET 52971		10.0
		2	11/17/83	TICKET 52972		10.0
		2	11/17/83	TICKET 52983		10.0
		2	11/17/83	TICKET 52984		10.0
		2	11/17/83	TICKET 52985		10.0
		2	11/17/83	TICKET 52986		10.0

RUN DATE: 25-JUL-88   JOB COST CUMULATIVE TRANSACTION REPORT   PAGE 36
JOB NAME :

COST-CODE	DESCRIPTION	TYPE	DATE	REFERENCE	$ AMOUNT	VOL AMOUNT
		2	11/17/83	TICKET 52987		10.0
		2	11/17/83	TICKET 52989		10.0
		2	11/17/83	TICKET 52990		10.0
		2	11/17/83	TICKET 53001		10.0
		2	11/17/83	TICKET 53002		10.0
		2	11/17/83	TICKET 53008		6.0
		2	11/17/83	TICKET 53017		4.0
		2	11/18/83	TICKET 53026		1.0
		2	11/18/83	TICKET 53031		10.0
		2	11/18/83	TICKET 53032		10.0
		2	11/18/83	TICKET 53033		10.0
		2	11/18/83	TICKET 53034		10.0
		2	11/18/83	TICKET 53035		10.0
		2	11/18/83	TICKET 53036		10.0
		2	11/18/83	TICKET 53037		10.0
		2	11/18/83	TICKET 53038		10.0
		2	11/18/83	TICKET 53039		10.0
		2	11/18/83	TICKET 53040		10.0
		2	11/18/83	TICKET 53046		10.0
		2	11/18/83	TICKET 53047		10.0
		2	11/18/83	TICKET 53048		10.0
		2	11/18/83	TICKET 53049		10.0
		2	11/18/83	TICKET 53054		10.0
		2	11/18/83	TICKET 53069		5.0
		2	11/21/83	TICKET 53070		10.0
		2	11/21/83	TICKET 53071		10.0
		2	11/21/83	TICKET 53072		10.0
		2	11/21/83	TICKET 53073		10.0
		2	11/21/83	TICKET 53074		10.0
		2	11/21/83	TICKET 53075		10.0
		2	11/21/83	TICKET 53076		10.0
		2	11/21/83	TICKET 53077		10.0
		2	11/21/83	TICKET 53078		10.0
		2	11/21/83	TICKET 53079		10.0
		2	11/21/83	TICKET 53080		10.0
				SUB-TOTALS	0.00	416.0
				TOTALS	0.00	416.0
007-209	CONC TOWER - SLABS - 9TH FLOOR	2	11/22/83	TICKET 53099		10.0
		2	11/22/83	TICKET 53100		10.0
		2	11/22/83	TICKET 53101		10.0
		2	11/22/83	TICKET 53102		10.0
		2	11/22/83	TICKET 53104		10.0
		2	11/22/83	TICKET 53105		10.0
		2	11/22/83	TICKET 53106		10.0
		2	11/22/83	TICKET 53107		10.0
		2	11/22/83	TICKET 53108		10.0
		2	11/22/83	TICKET 53109		10.0
		2	11/22/83	TICKET 53110		10.0
		2	11/22/83	TICKET 53111		10.0

44 ENTRIES FOR THIS TYPE
44 ENTRIES FOR THIS COST CODE

RUN DATE: 25-JUL-88　　　　　　　　JOB COST CUMULATIVE TRANSACTION REPORT　　　　　　　　PAGE 54
JOB NAME :

COST-CODE	DESCRIPTION	TYPE	DATE	REFERENCE	$ AMOUNT	VOL AMOUNT
		2	03/05/84	2 TICKET 52340		10.0
		2	03/05/84	2 TICKET 52341		10.0
		2	03/05/84	2 TICKET 52342		10.0
		2	03/05/84	2 TICKET 52343		10.0
		2	03/05/84	2 TICKET 52344		10.0
		2	03/05/84	2 TICKET 52345		10.0
		2	03/05/84	2 TICKET 52346		10.0
		2	03/05/84	2 TICKET 52350		10.0
		2	03/05/84	2 TICKET 52351		10.0
		2	03/05/84	2 TICKET 52356		10.0
		2	03/05/84	2 TICKET 52359 - SPLIT		3.0
		2	03/06/84	2 TICKET 52381		10.0
		2	03/06/84	2 TICKET 52382		10.0
		2	03/06/84	2 TICKET 52383		10.0
		2	03/06/84	2 TICKET 52384		10.0
		2	03/06/84	2 TICKET 52385		10.0
		2	03/06/84	2 TICKET 52388		10.0
		2	03/06/84	2 TICKET 52389		10.0
		2	03/06/84	2 TICKET 52390		10.0
		2	03/06/84	2 TICKET 52391		10.0
		2	03/06/84	2 TICKET 52392		10.0
		2	03/06/84	2 TICKET 52395		10.0
		2	03/06/84	2 TICKET 52396		10.0
		2	03/06/84	2 TICKET 52397		10.0
		2	03/06/84	2 TICKET 52398		10.0
		2	03/06/84	2 TICKET 52401		10.0
		2	03/06/84	2 TICKET 52404 - SPLIT		2.0
		2	03/07/84	2 TICKET 52410		10.0
		2	03/07/84	2 TICKET 52411		10.0
		2	03/07/84	2 TICKET 52412		10.0
		2	03/07/84	2 TICKET 52413		10.0
		2	03/07/84	2 TICKET 52414		10.0
		2	03/07/84	2 TICKET 52427		10.0
		2	03/07/84	2 TICKET 52428		10.0
		2	03/07/84	2 TICKET 52429		10.0
		2	03/07/84	2 TICKET 52430		10.0
		2	03/07/84	2 TICKET 52431		10.0
		2	03/07/84	2 TICKET 52433		10.0
		2	03/07/84	2 TICKET 52434		10.0
		2	03/07/84	2 TICKET 52441		10.0
		2	03/07/84	2 TICKET 52442 - SPLIT		8.0
				SUB-TOTALS	0.00	423.0
				TOTALS	0.00	423.0

****

007-229　CONC TOWER SLABS 29TH FLOOR

		2	03/08/84	2 TICKET 52473		10.0
		2	03/08/84	2 TICKET 52474		10.0
		2	03/08/84	2 TICKET 52475		10.0
		2	03/08/84	2 TICKET 52476		10.0
		2	03/08/84	2 TICKET 52477		10.0
		2	03/08/84	2 TICKET 52478		10.0

44 ENTRIES FOR THIS TYPE
44 ENTRIES FOR THIS COST CODE

```
RUN DATE: 25-JUL-88                    JOB COST CUMULATIVE TRANSACTION REPORT                        PAGE 55
JOB NAME :

COST-CODE  DESCRIPTION           TYPE  DATE      REFERENCE              $ AMOUNT    VOL AMOUNT

                                   2   03/08/84  TICKET 52479                           10.0
                                   2   03/08/84  TICKET 52480                           10.0
                                   2   03/08/84  TICKET 52481                           10.0
                                   2   03/08/84  TICKET 52482                           10.0
                                   2   03/08/84  TICKET 52487                           10.0
                                   2   03/08/84  TICKET 52488                           10.0
                                   2   03/08/84  TICKET 52490                           10.0
                                   2   03/08/84  TICKET 52491                            4.0
                                   2   03/09/84  TICKET 52500 - SPLIT                  10.0
                                   2   03/09/84  TICKET 52501                           10.0
                                   2   03/09/84  TICKET 52502                           10.0
                                   2   03/09/84  TICKET 52503                           10.0
                                   2   03/09/84  TICKET 52504                           10.0
                                   2   03/09/84  TICKET 52510                           10.0
                                   2   03/09/84  TICKET 52511                           10.0
                                   2   03/09/84  TICKET 52512                           10.0
                                   2   03/09/84  TICKET 52513                           10.0
                                   2   03/09/84  TICKET 52514                           10.0
                                   2   03/09/84  TICKET 52520                           10.0
                                   2   03/09/84  TICKET 52521                           10.0
                                   2   03/09/84  TICKET 52522                           10.0
                                   2   03/09/84  TICKET 52523                           10.0
                                   2   03/09/84  TICKET 52525                           10.0
                                   2   03/13/84  TICKET 52602                           10.0
                                   2   03/13/84  TICKET 52603                           10.0
                                   2   03/13/84  TICKET 52604                           10.0
                                   2   03/13/84  TICKET 52605                           10.0
                                   2   03/13/84  TICKET 52606                           10.0
                                   2   03/13/84  TICKET 52612                           10.0
                                   2   03/13/84  TICKET 52613                           10.0
                                   2   03/13/84  TICKET 52614                           10.0
                                   2   03/13/84  TICKET 52615                           10.0
                                   2   03/13/84  TICKET 52616                           10.0
                                   2   03/13/84  TICKET 52618                           10.0
                                   2   03/13/84  TICKET 52619                           10.0
                                   2   03/13/84  TICKET 52620                           10.0
                                   2   03/13/84  TICKET 52621                           10.0
                                                     SUB-TOTALS          0.00         424.0
                                                     TOTALS              0.00         424.0
                                   ****************************************************************
0007-230   CONC TOWER - SLABS - TOWER
                                   2   03/15/84  TICKET 52658                           10.0
                                   2   03/15/84  TICKET 52659                           10.0
                                   2   03/15/84  TICKET 52660                           10.0
                                   2   03/15/84  TICKET 52661                           10.0
                                   2   03/15/84  TICKET 52662                           10.0
                                   2   03/15/84  TICKET 52663                           10.0
                                   2   03/15/84  TICKET 52664                           10.0
                                   2   03/15/84  TICKET 52665                           10.0
                                   2   03/15/84  TICKET 52666                           10.0
                                   2   03/15/84  TICKET 52667                           10.0

43 ENTRIES FOR THIS TYPE
43 ENTRIES FOR THIS COST CODE
```

RUN DATE: 25-JUL-88   JOB COST CUMULATIVE TRANSACTION REPORT   PAGE 56

JOB NAME :

COST-CODE	DESCRIPTION	TYPE	DATE	REFERENCE	$ AMOUNT	VOL AMOUNT
		2	03/15/84	2 TICKET 52669		10.0
		2	03/15/84	2 TICKET 52670		10.0
		2	03/15/84	2 TICKET 52671		5.0
		2	03/15/84	2 TICKET 52673		10.0
		2	03/19/84	2 TICKET 52710		10.0
		2	03/19/84	2 TICKET 52711		10.0
		2	03/19/84	2 TICKET 52712		10.0
		2	03/19/84	2 TICKET 52713		10.0
		2	03/19/84	2 TICKET 52714		10.0
		2	03/19/84	2 TICKET 52717		10.0
		2	03/19/84	2 TICKET 52718		10.0
		2	03/19/84	2 TICKET 52719		10.0
		2	03/19/84	2 TICKET 52720		10.0
		2	03/19/84	2 TICKET 52721		10.0
		2	03/19/84	2 TICKET 52722		10.0
		2	03/19/84	2 TICKET 52723		10.0
		2	03/19/84	2 TICKET 52724		10.0
		2	03/19/84	2 TICKET 52725		10.0
		2	03/19/84	2 TICKET 52731		10.0
		2	03/19/84	2 TICKET 52734 - SPLIT		2.0
		2	03/21/84	2 TICKET 52757		10.0
		2	03/21/84	2 TICKET 52758		10.0
		2	03/21/84	2 TICKET 52759		10.0
		2	03/21/84	2 TICKET 52760		10.0
		2	03/21/84	2 TICKET 52761		10.0
		2	03/21/84	2 TICKET 52765		10.0
		2	03/21/84	2 TICKET 52766		10.0
		2	03/21/84	2 TICKET 52767		10.0
		2	03/21/84	2 TICKET 52768		10.0
		2	03/21/84	2 TICKET 52769		10.0
		2	03/21/84	2 TICKET 52766		10.0
		2	03/21/84	2 TICKET 52777		10.0
		2	03/21/84	2 TICKET 52779		10.0
		2	03/21/84	2 TICKET 52780 - SPLIT		8.0
				SUB-TOTALS	0.00	425.0

44 ENTRIES FOR THIS TYPE

 0 ENTRIES FOR THIS TYPE   SUB-TOTALS   0.00   0.0
44 ENTRIES FOR THIS COST CODE   TOTALS   0.00 425.0
007-231 CONC TOWER - SLABS - P/H LOWER
****

| | | 1 | 05/10/84 | 1 MIRON CK 551 | 110.25 | |
| | | | | SUB-TOTALS | 110.25 | 0.0 |

1 ENTRY FOR THIS TYPE

		2	03/23/84	2 TICKET 50830		10.0
		2	03/23/84	2 TICKET 50831		10.0
		2	03/23/84	2 TICKET 50832		10.0
		2	03/23/84	2 TICKET 50833		10.0
		2	03/23/84	2 TICKET 50834		10.0
		2	03/23/84	2 TICKET 50847		10.0
		2	03/23/84	2 TICKET 50848		10.0
		2	03/23/84	2 TICKET 50849		10.0

RUN DATE: 25-JUL-88　　　　　　　　　JOB COST CUMULATIVE TRANSACTION REPORT　　　　　　　　　PAGE 57
JOB NAME :

COST-CODE	DESCRIPTION	TYPE	DATE	REFERENCE	$ AMOUNT	VOL AMOUNT
		2	03/23/84	2 TICKET 50850		10.0
		2	03/23/84	2 TICKET 50851		10.0
		2	03/23/84	2 TICKET 50854		10.0
		2	03/23/84	2 TICKET 50855		10.0
		2	03/23/84	2 TICKET 50857		10.0
		2	03/23/84	2 TICKET 50858		10.0
		2	03/23/84	2 TICKET 50859 - SPLIT		2.0
		2	03/27/84	2 TICKET 50878		10.0
		2	03/27/84	2 TICKET 50879		10.0
		2	03/27/84	2 TICKET 50880		10.0
		2	03/27/84	2 TICKET 50881		10.0
		2	03/27/84	2 TICKET 50882		10.0
		2	03/27/84	2 TICKET 50886		10.0
		2	03/27/84	2 TICKET 50887		10.0
		2	03/27/84	2 TICKET 50888		10.0
		2	03/27/84	2 TICKET 50889		10.0
		2	03/27/84	2 TICKET 50890		10.0
		2	03/27/84	2 TICKET 50892		10.0
		2	03/27/84	2 TICKET 50893		10.0
		2	03/27/84	2 TICKET 50894		10.0
		2	03/27/84	2 TICKET 50895		10.0
		2	03/27/84	2 TICKET 50896		10.0
		2	03/27/84	2 TICKET 50900		10.0
		2	04/02/84	2 TICKET 50969		10.0
		2	04/02/84	2 TICKET 50970		10.0
		2	04/02/84	2 TICKET 50971		10.0
		2	04/02/84	2 TICKET 50972		10.0
		2	04/02/84	2 TICKET 50973		10.0
		2	04/02/84	2 TICKET 50976		10.0
		2	04/02/84	2 TICKET 50977		10.0
		2	04/02/84	2 TICKET 50978		10.0
		2	04/02/84	2 TICKET 50979		10.0
		2	04/02/84	2 TICKET 50980		10.0
		2	04/02/84	2 TICKET 50982		10.0
		2	04/02/84	2 TICKET 50983		10.0
		2	04/02/84	2 TICKET 50984		10.0
		2	04/02/84	2 TICKET 50985		10.0
		2	04/02/84	2 TICKET 50988 - SPLIT		1.0
				SUB-TOTALS	0.00	443.0
				TOTALS	110.25	443.0
				**************		
007-232	CONC TOWER - SLABS - P/H UPPER	2	04/06/84	2 TICKET 51052		10.0
		2	04/06/84	2 TICKET 51053		10.0
		2	04/06/84	2 TICKET 51054		10.0
		2	04/06/84	2 TICKET 51055		10.0
		2	04/06/84	2 TICKET 51056		10.0
		2	04/06/84	2 TICKET 51057		10.0
		2	04/06/84	2 TICKET 51058		10.0
		2	04/06/84	2 TICKET 51059		10.0
		2	04/06/84	2 TICKET 51060		10.0

46 ENTRIES FOR THIS TYPE
47 ENTRIES FOR THIS COST CODE

RUN DATE: 25-JUL-88 JOB COST CUMULATIVE TRANSACTION REPORT PAGE 58

JOB NAME :

COST-CODE DESCRIPTION	TYPE	DATE	REFERENCE		$ AMOUNT	VOL AMOUNT
	2	04/05/84	TICKET 51061			10.0
	2	04/05/84	TICKET 51068			10.0
	2	04/05/84	TICKET 51069			10.0
	2	04/05/84	TICKET 51070			10.0
	2	04/05/84	TICKET 51071			10.0
	2	04/05/84	TICKET 51073			10.0
	2	04/05/84	TICKET 51074			10.0
	2	04/05/84	TICKET 51075			10.0
	2	04/05/84	TICKET 51076			10.0
	2	04/05/84	TICKET 51077			10.0
	2	04/05/84	TICKET 51078			10.0
	2	04/05/84	TICKET 51080			10.0
	2	04/05/84	TICKET 51081			10.0
	2	04/05/84	TICKET 51082			10.0
	2	04/10/84	TICKET 51114			10.0
	2	04/10/84	TICKET 51115			10.0
	2	04/10/84	TICKET 51116			10.0
	2	04/10/84	TICKET 51117			10.0
	2	04/10/84	TICKET 51118			10.0
	2	04/10/84	TICKET 51120			10.0
	2	04/10/84	TICKET 51121			10.0
	2	04/10/84	TICKET 51122			10.0
	2	04/10/84	TICKET 51123			10.0
	2	04/10/84	TICKET 51124			10.0
	2	04/10/84	TICKET 51125			10.0
	2	04/10/84	TICKET 51126			10.0
	2	04/10/84	TICKET 51127			10.0
	2	04/10/84	TICKET 51128			10.0
	2	04/10/84	TICKET 51129			10.0
	2	04/10/84	TICKET 51130			10.0
	2	04/10/84	TICKET 51131			10.0
	2	04/10/84	TICKET 51132			10.0
	2	04/10/84	TICKET 51133			10.0
	2	04/10/84	TICKET 51134			10.0
	2	04/10/84	TICKET 51135			10.0
	2	04/10/84	TICKET 51139	- SPLIT		4.0
	2	04/11/84	TICKET 51151			10.0
	2	04/12/84	TICKET 51158			10.0
	2	04/12/84	TICKET 51159			10.0
	2	04/12/84	TICKET 51160			10.0
	2	04/12/84	TICKET 51161			10.0
	2	04/12/84	TICKET 51162			10.0
	2	04/12/84	TICKET 51163			10.0
	2	04/12/84	TICKET 51164			10.0
	2	04/12/84	TICKET 51165			10.0
	2	04/12/84	TICKET 51166			10.0
	2	04/12/84	TICKET 51167			10.0
	2	04/12/84	TICKET 51172			10.0
	2	04/12/84	TICKET 51173			10.0

RUN DATE: 25-JUL-88   JOB COST CUMULATIVE TRANSACTION REPORT   PAGE 59
JOB NAME :

COST-CODE	DESCRIPTION	TYPE	DATE	REFERENCE	$ AMOUNT	VOL AMOUNT
		2	04/12/84	2 TICKET 51174		10.0
		2	04/12/84	2 TICKET 51175		10.0
		2	04/12/84	2 TICKET 51176		10.0
		2	04/12/84	2 TICKET 51178		10.0
		2	04/12/84	2 TICKET 51179		10.0
		2	04/12/84	2 TICKET 51180		10.0
		2	04/12/84	2 TICKET 51181		10.0
		2	04/12/84	2 TICKET 51182		10.0
		2	04/12/84	2 TICKET 51185		10.0
		2	04/12/84	2 TICKET 51186		10.0
		2	04/12/84	2 TICKET 51187		10.0
		2	04/12/84	2 TICKET 51188		10.0
		2	04/12/84	2 TICKET 51189		10.0
		2	04/12/84	2 TICKET 51190		10.0
		2	04/12/84	2 TICKET 51191		10.0
		2	04/12/84	2 TICKET 51193		10.0
		2	04/12/84	2 TICKET 51194		10.0
		2	04/12/84	2 TICKET 51195		10.0
				SUB-TOTALS	0.00	774.0
				TOTALS	0.00	774.0
				************	********	********
		2	04/19/84	2 TICKET 51261		10.0
		2	04/19/84	2 TICKET 51262		10.0
		2	04/19/84	2 TICKET 51263		10.0
		2	04/19/84	2 TICKET 51264		10.0
		2	04/19/84	2 TICKET 51265		10.0
		2	04/19/84	2 TICKET 51266		10.0
		2	04/19/84	2 TICKET 51267		10.0
		2	04/19/84	2 TICKET 51277		10.0
		2	04/19/84	2 TICKET 51279		10.0
		2	04/19/84	2 TICKET 51281 - SPLIT		8.0
		2	04/24/84	2 TICKET 51335		10.0
		2	04/24/84	2 TICKET 51336		10.0
		2	04/24/84	2 TICKET 51337		10.0
		2	04/24/84	2 TICKET 51338		10.0
		2	04/24/84	2 TICKET 51339		10.0
		2	04/24/84	2 TICKET 51341		10.0
		2	04/24/84	2 TICKET 51342		10.0
		2	04/24/84	2 TICKET 51343		10.0
		2	04/24/84	2 TICKET 51345		10.0
		2	04/24/84	2 TICKET 51347		10.0
		2	04/26/84	2 TICKET 51399 - SPLIT		5.0
		2	04/27/84	2 TICKET 51424		10.0
		2	04/30/84	2 TICKET 51444		10.0
		2	04/30/84	2 TICKET 51445		10.0
		2	04/30/84	2 TICKET 51446		10.0
		2	04/30/84	2 TICKET 51447		10.0
		2	04/30/84	2 TICKET 51448		10.0
		2	04/30/84	2 TICKET 51450		10.0
		2	04/30/84	2 TICKET 51451		10.0

78 ENTRIES FOR THIS TYPE
78 ENTRIES FOR THIS COST CODE
007-233 CONC TOWER - SLABS - P/H ROOF

RUN DATE: 25-JUL-88   JOB COST CUMULATIVE TRANSACTION REPORT   PAGE 60
JOB NAME :

COST-CODE	DESCRIPTION	TYPE	DATE	REFERENCE	$ AMOUNT	VOL AMOUNT
		2	04/30/84	TICKET 51452		10.0
		2	04/30/84	TICKET 51453		10.0
		2	04/30/84	TICKET 51454		10.0
		2	04/30/84	TICKET 51455		10.0
		2	04/30/84	TICKET 51458		10.0
		2	05/02/84	TICKET 51518		8.0
		2	05/02/84	TICKET 51525 - SPLIT		7.0
		2	05/03/84	TICKET 51548 -SPLIT		10.0
		2	05/04/84	TICKET 51582		4.0
		2	05/07/84	TICKET 51601 - SPLIT		5.0
		2	05/07/84	TICKET 51605		5.0
		2	05/07/84	TICKET 51609		3.0
		2	05/11/84	TICKET 51700 - SPLIT		
				SUB-TOTALS	0.00	385.0
				TOTALS	0.00	385.0
		2	05/02/84	TICKET 51525 - SPLIT		2.0
		2	05/02/84	TICKET 51531		3.0
		2	05/03/84	TICKET 51548 - SPLIT		3.0
		2	05/03/84	TICKET 51550		10.0
		2	05/03/84	TICKET 51557		10.0
		2	05/03/84	TICKET 51560		10.0
		2	05/03/84	TICKET 51561		10.0
		2	05/11/84	TICKET 51682		10.0
		2	05/11/84	TICKET 51683		10.0
		2	05/11/84	TICKET 51684		10.0
		2	05/11/84	TICKET 51685		10.0
		2	05/11/84	TICKET 51686		10.0
		2	05/11/84	TICKET 51687		5.0
		2	05/11/84	TICKET 51700 - SPLIT		10.0
		2	05/21/84	TICKET 53849		6.0
		2	05/21/84	TICKET 53853 - SPLIT		3.0
		2	07/10/84	54726		1.5
		2	07/27/84	56283		5.0
		2	08/06/84	56410 - SPLIT		
		2	08/16/84	56635		
				SUB-TOTALS	0.00	142.0
				TOTALS	0.00	142.0
		2	04/09/84	TICKET 51092		8.0
		2	12/21/84	59362		8.5
		2	12/21/84	59344		10.0
		2	12/28/84	59451		10.0
				SUB-TOTALS	0.00	36.5
				TOTALS	0.00	36.5
		2	03/27/84	TICKET 50899		10.0
		2	03/27/84	TICKET 50901		10.0

007-234 42 ENTRIES FOR THIS TYPE
         42 ENTRIES FOR THIS COST CODE
         CONC TOWER - SLABS - MACHINE

007-236 20 ENTRIES FOR THIS TYPE
         20 ENTRIES FOR THIS COST CODE
         CONC TOWER - 4" TOPPING 3RD FL

007-237 4 ENTRIES FOR THIS TYPE
         4 ENTRIES FOR THIS COST CODE
         CONC TOWER - 4" TOPPING 4TH FL

RUN DATE: 25-JUL-88　　　　　　　　　　JOB COST CUMULATIVE TRANSACTION REPORT　　　　　　　　　　PAGE 61
JOB NAME :

COST-CODE	DESCRIPTION	TYPE	DATE	REFERENCE	$ AMOUNT	VOL AMOUNT
		2	03/27/84	2 TICKET 50904		10.0
		2	03/27/84	2 TICKET 50905		10.0
		2	03/27/84	2 TICKET 50909		10.0
		2	03/30/84	2 TICKET 50948		8.0
		2	04/09/84	2 TICKET 51104		8.0
		2	10/17/84	2 57923		10.0
		2	12/28/84	2 59454 - SPLIT		5.0
		2	03/15/85	2 22753		10.0
		2	03/29/85	2 19023		6.0
		2	03/29/85	2 19008		10.0
				SUB-TOTALS	0.00	107.0
				TOTALS	0.00	107.0
007-238	12 ENTRIES FOR THIS TYPE					
	12 ENTRIES FOR THIS COST CODE					
	CONC TOWER - 4" TOPPING P/H					
		2	09/18/84	2 57362		10.0
		2	10/01/84	2 57575		5.0
		2	10/02/84	2 57595		2.0
		2	10/08/84	2 57706		5.0
		2	10/09/84	2 57724		5.0
		2	10/10/84	2 57737		5.0
		2	10/10/84	2 57803		5.0
		2	10/11/84	2 57804		5.0
		2	10/11/84	2 57813		5.0
		2	10/12/84	2 58330		5.0
		2	10/15/84	2 57870 - SPLIT		4.0
		2	10/15/84	2 57871		5.0
		2	10/16/84	2 57896		5.0
		2	10/16/84	2 57899		5.0
		2	10/16/84	2 57901 - SPLIT		4.0
		2	10/19/84	2 58003		5.0
		2	10/19/84	2 58065		5.0
		2	10/23/84	2 58019		3.0
		2	10/24/84	2 58077		5.0
		2	10/24/84	2 58063		5.0
		2	10/24/84	2 58098		5.0
		2	10/26/84	2 58131 - SPLIT		8.0
		2	10/30/84	2 58182 - SPLIT		3.0
				SUB-TOTALS	0.00	114.0
				TOTALS	0.00	114.0
007-239	23 ENTRIES FOR THIS TYPE					
	23 ENTRIES FOR THIS COST CODE					
	CONC TOWER - 5" L/W TOPPING					
		2	08/06/84	2 56404		10.0
		2	08/06/84	2 56405		10.0
		2	08/06/84	2 56406		10.0
		2	08/06/84	2 56407		10.0
		2	08/06/84	2 56409		10.0
		2	08/06/84	2 56410 - SPLIT		8.5
				SUB-TOTALS	0.00	58.5
				TOTALS	0.00	58.5
007-240	6 ENTRIES FOR THIS TYPE					
	6 ENTRIES FOR THIS COST CODE					
	HOIST PAD					

```
RUN DATE: 25-JUL-88                    JOB COST CUMULATIVE TRANSACTION REPORT                        PAGE  62
JOB NAME :

COST-CODE  DESCRIPTION                  TYPE  DATE      REFERENCE                    $ AMOUNT      VOL AMOUNT

                                          2  11/12/83  2  LONESTAR INVOICE #50039                      14.0
                                          2  11/22/83  2  TICKET 53118 - SPLIT                          2.0
                                          2  11/28/83  2  TICKET 53189 - SPLIT                          4.0
                                          2  11/29/83  2  TICKET 53219                                  6.0
                                                         SUB-TOTALS                     0.00           26.0
           4 ENTRIES FOR THIS TYPE                       TOTALS                         0.00           26.0
           4 ENTRIES FOR THIS COST CODE  *******************************************************************
007-310    CONC TOWER - BEAMS - HIGH ROOF
                                          2  05/16/84  2  TICKET 51779                                 10.0
                                          2  05/16/84  2  TICKET 51780                                  5.0
                                          2  05/21/84  2  TICKET 53850                                 10.0
                                          2  05/21/84  2  TICKET 53853 - SPLIT                          2.0
                                                         SUB-TOTALS                     0.00           27.0
           4 ENTRIES FOR THIS TYPE                       TOTALS                         0.00           27.0
           4 ENTRIES FOR THIS COST CODE
007-311    CONC TOWER - BEAMS - PARAPET
                                          2  05/25/84  2  TICKET 52967                                 10.0
                                          2  05/25/84  2  TICKET 53968                                 10.0
                                          2  05/30/84  2  TICKET 53987                                  5.0
                                          2  06/01/84  2  TICKET 01327 - SPLIT                          5.0
                                                         SUB-TOTALS                     0.00           30.0
           4 ENTRIES FOR THIS TYPE       *******************************************************************
           0 ENTRIES FOR THIS TYPE                       SUB-TOTALS                     0.00            0.0
           4 ENTRIES FOR THIS COST CODE                  TOTALS                         0.00           30.0
008-101    DEWATER - SYSTEM INSTALLATION
                                          1  08/04/83  1  BC BUILDERS                  20.58
                                          1  08/04/83  1  LONESTAR FLA                147.76
                                          1  09/21/83  1                              100.00
                                          1  09/21/83  1                               50.00
                                                         SUB-TOTALS                   318.34
           4 ENTRIES FOR THIS TYPE
                                          3  08/04/83  3  CENTRAL FLA                 504.00
                                          3  08/04/83  3  CENTRAL FLA                 945.00
                                                         SUB-TOTALS                 1,449.00
           2 ENTRIES FOR THIS TYPE                       TOTALS                     1,767.34
           6 ENTRIES FOR THIS COST CODE  *******************************************************************
008-102    DEWATER - SYSTEM REMOVAL
                                          3  09/12/83  3  CENTRAL FLA                 108.00
                                                         SUB-TOTALS                   108.00
           1 ENTRY FOR THIS TYPE                         TOTALS                       108.00
           1 ENTRY FOR THIS COST CODE    *******************************************************************
008-103    DEWATER - 2" PUMPS
                                          1  08/02/83  1  PETTY CASH                   15.00
                                                         SUB-TOTALS                    15.00
           1 ENTRY FOR THIS TYPE
                                          3  06/29/83  3  PETTY CASH                   15.00
                                          3  07/06/83  3  PETTY CASH                   12.00
                                          3  08/04/83  3  BLANCHARD                    36.00
                                                                                      679.78
```

250

RUN DATE: 25-JUL-88
JOB COST CUMULATIVE TRANSACTION REPORT

JOB NAME : 004 WILLIAMS ISLAND

COST-CODE	DESCRIPTION	TYPE	DATE	REFERENCE	$ AMOUNT	VOL AMOUNT
		3	09/12/83	3 BLANCHARD	135.82	
		3	03/10/84	3 BLANCHARD - CK 289	751.80	
	5 ENTRIES FOR THIS TYPE			SUB-TOTALS	1,615.40	0.0
	0 ENTRIES FOR THIS TYPE			SUB-TOTALS	0.00	0.0
	6 ENTRIES FOR THIS COST CODE			TOTALS	1,630.40	0.0
008-104	DEWATER - MAIN PUMP	1	08/04/83	1 LONESTAR	150.00	
	1 ENTRY FOR THIS TYPE			SUB-TOTALS	150.00	0.0
		3	10/05/83	3 IRRIGATOR PUMP	2,484.83	
	1 ENTRY FOR THIS TYPE			SUB-TOTALS	2,484.83	0.0
	2 ENTRIES FOR THIS COST CODE			TOTALS	2,634.83	0.0
009-101	EARTHWORK - RAFT EXCAVATE	3	08/04/83	3 CENTRAL FLA	6,131.50	
		3	08/04/83	3 CENTRAL FLA	315.00	
		3	09/12/83	3 CENTRAL FLA	652.50	
	3 ENTRIES FOR THIS TYPE			SUB-TOTALS	7,099.00	0.0
	3 ENTRIES FOR THIS COST CODE			TOTALS	7,099.00	0.0
009-102	EARTHWORK - RAFT BACKFILL TO	3	09/12/83	3 CENTRAL FLA EQUIP	3,145.50	
		3	10/10/83	3 CENTRAL FLA EQUIP	3,793.50	
	2 ENTRIES FOR THIS TYPE			SUB-TOTALS	6,939.00	0.0
	2 ENTRIES FOR THIS COST CODE			TOTALS	6,939.00	0.0
009-103	EARTHWORK - T/H EXCAVATION	3	09/12/83	3 CENTRAL FLA EQUIP	108.00	
		3	10/10/83	3 CENTRAL FLA EQUIP	1,026.00	
		3	01/10/84	3 CENTRAL FLA EQUIP - C	180.00	
	3 ENTRIES FOR THIS TYPE			SUB-TOTALS	1,314.00	0.0
	3 ENTRIES FOR THIS COST CODE			TOTALS	1,314.00	0.0
009-104	EARTHWORK - T/H BACKFILL TO	3	10/10/83	3 CENTRAL FLA EQUIP	3,658.50	
		3	11/10/83	3 BOYS PLUMBING	134.00	
	2 ENTRIES FOR THIS TYPE			SUB-TOTALS	3,792.50	0.0
	2 ENTRIES FOR THIS COST CODE			TOTALS	3,792.50	0.0
009-105	EARTHWORK - GARAGE EXCAVATE	3	09/12/83	3 CENTRAL FLA EQUIP	180.00	
		3	10/10/83	3 CENTRAL FLA EQUIP	1,404.00	
		3	01/10/84	3 CENTRAL FLA EQUIP - C	1,728.00	
		3	02/10/84	3 CENTRAL FLA - CHECK #	1,980.00	
		3	04/10/84	3 CENTRAL FLA CK 463	162.50	
		3	07/10/84	3 CENTRAL FLA CK 751	1,615.00	
		3	08/10/84	3 CENT FLA CK 831	950.00	
		3	09/10/84	3 CENTRAL FLA CK 913	612.00	

PAGE 6.3

RUN DATE: 25-JUL-88   JOB COST CUMULATIVE TRANSACTION REPORT   PAGE 64
JOB NAME : 004 WILLIAMS ISLAND

COST-CODE	DESCRIPTION	TYPE	DATE	REFERENCE	$ AMOUNT	VOL AMOUNT
009-106	EARTHWORK - GARAGE BACKFILL TO					
		3	11/10/83	SNEAD-CANIPE	378.00	
		3	12/09/83	CENTRAL FLA EQUIPMENT	280.00	
		3	01/10/84	CENTRAL FLA EQUIP - C	360.00	
		3	02/10/84	CENTRAL FLA - CHECK #	3,147.50	
		3	03/10/84	CENTRAL FLA - CK 291	3,838.00	
		3	07/10/84	CENTRAL FLA CK 751	266.00	
		3	08/10/84	CENT FL CK 831	1,703.00	
		3	08/10/84	SNEAD CK 839	925.00	
	8 ENTRIES FOR THIS TYPE			SUB-TOTALS	10,897.50	0.0
	8 ENTRIES FOR THIS COST CODE			TOTALS	10,897.50	0.0
009-107	EARTHWORK - GARAGE 2ND LEVEL					
		3	05/10/84	CENTRAL EQUIP CK 542	594.00	
		3	02/08/85	CENT FL CK 1460	305.00	
		3	02/08/85	METRO TRUCKING CK 147	517.00	
		3	03/10/85	SNEAD-CANIPE CK 1527	1,757.00	
		3	04/08/85	SNEAD CANIPE CK 1593	1,264.00	
		3	05/09/85	HARRIS BACKHOE CK 167	148.00	
		3	05/09/85	SNEAD-CANIPE CK 1658	711.00	
	7 ENTRIES FOR THIS TYPE			SUB-TOTALS	5,296.00	0.0
	7 ENTRIES FOR THIS COST CODE			TOTALS	5,296.00	0.0
009-108	EARTHWORK - POOL DECK - FILL					
		3	12/10/84	CENT FLA CK 1239	3,610.60	
		3	01/10/85	CENT FLA CK 1368	4,255.00	
		3	02/08/85	CENT FL CK 1460	770.00	
		3	02/08/85	CENT FL CK 1460	858.00	
		3	03/10/85	SNEAD-CANIPE CK 1527	1,991.00	
		3	04/04/85	LATTIMORE CK 4305	3,916.00	
		3	04/08/85	SNEAD CANIPE CK 1593	1,784.00	
	7 ENTRIES FOR THIS TYPE			SUB-TOTALS	17,184.60	0.0
	7 ENTRIES FOR THIS COST CODE			TOTALS	17,184.60	0.0
009-109	EARTHWORK - PLANTS & SITE -					
		1	04/15/85	JIM TATUM CK 4334	1,250.00	
		1	05/03/85	LATTIMORE CK 4392	1,100.00	
		1	06/10/85	LATTIMORE CK 1680	178.20	
	3 ENTRIES FOR THIS TYPE			SUB-TOTALS	2,528.20	0.0
		3	09/10/84	CENTRAL FLA CK 913	1,461.00	
		3	09/15/84	CONDUIT & FDTN CK 100	4,416.00	
		3	01/10/85	CENT FLA CK 1368	300.00	
		3	02/08/85	CENT FLA CK 1460	891.00	
		3	02/08/85	METRO TRUCKING CK 147	2,244.00	
		3	03/10/85	SNEAD-CANIPE CK 1527	2,539.00	
		3	04/08/85	SNEAD CANIPE CK 1593	13,246.50	

RUN DATE: 25-JUL-88 JOB COST CUMULATIVE TRANSACTION REPORT PAGE 65

JOB NAME : 004 WILLIAMS ISLAND

COST-CODE	DESCRIPTION	TYPE	DATE	REFERENCE	$ AMOUNT	VOL AMOUNT
		3	04/08/85	METRO TRKG CK 1606	2,915.00	
		3	05/09/85	HARRIS BACKHOE CK 167	185.00	
		3	05/09/85	SNEAD-CANIPE CK 1658	2,300.50	
		3	05/22/85	P/C CK 4448	212.10	
		3	06/10/85	SNEAD-CANIPE CK 1683	3,633.50	
		3	07/10/85	CREST CONCRETE CK 172	1,390.00	
				SUB-TOTALS	31,733.60	0.0
				TOTALS	34,261.80	0.0
009-110	13 ENTRIES FOR THIS TYPE					
	16 ENTRIES FOR THIS COST CODE					
	EARTHWORK - SITE EXCAVATE -					
		3	08/10/84	CENT FL CK 831	300.00	
		3	03/10/85	METRO TRKG CK 1537	1,008.00	
		3	04/08/85	SNEAD CANIPE CK 1593	344.00	
		3	06/10/85	SNEAD-CANIPE CK 1683		
				SUB-TOTALS	1,964.00	0.0
				TOTALS	1,964.00	0.0
009-111	4 ENTRIES FOR THIS TYPE					
	4 ENTRIES FOR THIS COST CODE					
	EARTHWORK - COMPACTIONS					
		3	09/12/83	BLANCHARD	326.92	
		3	02/10/84	BLANCHARD - CHECK #17	985.74	
		3	03/10/84	BLANCHARD - CK 289	367.34	
		3	03/10/84	SNEAD CANIPE CK 301	432.00	
		3	05/10/84	BLANCHARD CK 540	373.91	
		3	08/10/84	BLANCHARD CK 829	144.22	
		3	08/10/84	BLANCHARD CK 829	302.40	
		3	09/10/84	CENT FL CK 831	100.00	
		3	09/10/84	BLANCHARD CK 911	122.94	
		3	02/08/85	BLANCHARD CK 1459	687.07	
		3	04/08/85	BLANCHARD CK 1586	1,164.82	
		3	05/09/85	BLANCHARD CK 1651	666.44	
		3	06/10/85	BLANCHARD CK 1678	649.26	
				SUB-TOTALS	6,373.06	
		9	09/13/83	PETTY CASH	22.00	
		9	10/10/83	BLANCHARD	1,002.91	
		9	10/10/83	PURDY SOIL	45.00	
				SUB-TOTALS	1,069.91	0.0
				TOTALS	7,442.97	0.0
	13 ENTRIES FOR THIS TYPE					
	3 ENTRIES FOR THIS TYPE					
	0 ENTRIES FOR THIS TYPE					
	16 ENTRIES FOR THIS COST CODE					
009-113	3/4" DRAIN ROCK					
		1	12/10/84	METRO CK 1254	1,254.46	
		1	01/10/85	METRO TRK CK 1384	668.89	
		1	05/09/85	METRO TRUCK CK 1669	143.37	
				SUB-TOTALS	2,066.72	
		3	02/08/85	METRO TRUCKING CK 147	1,163.83	
		3	03/10/85	METRO TRKG CK 1537	438.80	
				SUB-TOTALS	1,602.63	0.0
	3 ENTRIES FOR THIS TYPE					
	2 ENTRIES FOR THIS TYPE					

RUN DATE: 25-JUL-88
JOB NAME :

JOB COST CUMULATIVE TRANSACTION REPORT

PAGE 66

COST-CODE	DESCRIPTION		TYPE	DATE	REFERENCE	$ AMOUNT	VOL AMOUNT
009-114	0 ENTRIES FOR THIS TYPE 5 ENTRIES FOR THIS COST CODE MIRAFI - DRAINAGE MATERIAL				SUB-TOTALS TOTALS	0.00 3,669.35	0.0 0.0
			1	11/16/84	P/C CK 4025	384.00	
			1	12/10/84	ENG SYS CK 1255	1,153.50	
			1	01/10/85	ENG SYS CK 185	384.50	
			1	02/08/85	ENGINEERED SYS. CK 14	1,155.50	
			1	03/10/85	ENGINEERED SYS CK 153	1,153.50	
010-101	5 ENTRIES FOR THIS TYPE 5 ENTRIES FOR THIS COST CODE GEN COND - FIELD OFF & SHEDS				SUB-TOTALS TOTALS	4,231.00 4,231.00	0.0 0.0
			1	09/22/83	PETTY CASH	115.50	
			1	04/10/84	MIRON CK 474	27.29	
	2 ENTRIES FOR THIS TYPE				SUB-TOTALS	142.79	
			9	08/04/83	AZCO	1,197.00	
			9	09/09/83	AZCO	378.00	
			9	10/10/83	AZCO	378.00	
			9	11/10/83	AZCO	378.00	
			9	12/09/83	AZCO EQUIPMENT	378.00	
			9	01/10/84	AZCO EQUIPMENT - CK 3	378.00	
			9	02/10/84	AZCO - CHECK #191	378.00	
			9	03/10/84	AZCO - CK 305	378.00	
			9	04/10/84	AZCO CK 475	378.00	
			9	05/10/84	AZCO CK 552	378.00	
			9	06/08/84	AZCO CK 691	378.00	
			9	08/10/84	AZCO CK 842	157.50	
	12 ENTRIES FOR THIS TYPE				SUB-TOTALS	5,134.50	
010-102	0 ENTRIES FOR THIS TYPE 14 ENTRIES FOR THIS COST CODE GEN COND - OFF EQUIP & SUP				SUB-TOTALS TOTALS	0.00 5,277.29	0.0 0.0
			1	06/24/83	PETTY CASH	72.58	
			1	06/29/83	PETTY CASH	241.49	
			1	06/29/83	PETTY CASH	29.21	
			1	07/06/83	PETTY CASH	16.74	
			1	07/15/83	PETTY CASH	15.39	
			1	07/15/83	PETTY CASH	8.04	
			1	08/02/83	PETTY CASH	44.44	
			1	08/04/83	BC BUILDERS	113.39	
			1	08/04/83	DECORA STEEL	655.20	
			1	08/04/83	MIRON	16.80	
			1	08/04/83	PAN AMER LUMBER	73.42	
			1	08/04/83	ZEE MEDICAL SVS	178.10	
			1	08/10/83	KEMRON	43.07	
			1	08/17/83	PETTY CASH	450.00	
			1	08/18/83	PETTY CASH	79.15	
			1	09/01/83	PETTY CASH	72.58	

RUN DATE: 25-JUL-88   JOB COST CUMULATIVE TRANSACTION REPORT   PAGE 70

JOB NAME :

COST-CODE	DESCRIPTION	TYPE	DATE	REFERENCE	$ AMOUNT	VOL AMOUNT
		1	07/10/85 1	T-SQUARE CK 1719	9.26	
		1	07/17/85 1	P/C CK 4584	15.86	
		1	07/23/85 1	P/C CK 4591	25.70	
		1	08/16/85 1	P/C SHEET 8/16/85	60.82	
		1	09/10/85 1	MOTOROLA CK 1772 FINA	55.75	
		1	10/04/85 1	P/C CK 4723	7.64	
				SUB-TOTALS	8,928.59	0.0
		9	05/10/84 9	SUPREME COPY CK 555	94.50	
		9	03/10/85 9	SUPREME CK 1530	59.33	
				SUB-TOTALS	153.83	0.0

175 ENTRIES FOR THIS TYPE

| | | | | SUB-TOTALS | 0.00 | 0.0 |
| | | | | TOTALS | 9,082.42 | 0.0 |

2 ENTRIES FOR THIS TYPE

****
0 ENTRIES FOR THIS TYPE
010-103 GEN COND - MOVE TO BUILDING
177 ENTRIES FOR THIS COST CODE

		1	07/10/84 1	MIRON CK 759	104.75	
				SUB-TOTALS	104.75	0.0
				TOTALS	104.75	0.0

****
1 ENTRY FOR THIS TYPE
010-104 GEN COND - TELEPHONE
1 ENTRY FOR THIS COST CODE

		1	08/16/85 1	P/C SHEET 8/16/85	1.05	
				SUB-TOTALS	1.05	0.0
		9	07/25/83 9	SO BELL	536.46	
		9	08/15/83 9	SO BELL	131.42	
		9	08/22/83 9	SO BELL	242.19	
		9	09/16/83 9	SO BELL	20.99	
		9	09/22/83 9	SO BELL	224.93	
		9	10/18/83 9	SOUTHERN BELL	20.99	
		9	10/26/83 9	PAY PHONE WMS ISLAND	175.46	
		9	11/16/83 9	WMS ISL PAY TELEPHONE	20.99	
		9	11/30/83 9	SOUTHERN BELL	295.79	
		9	12/29/83 9	SO. BELL CK 35	200.22	
		9	01/26/84 9	SO. BELL CK 3411	20.99	
		9	01/26/84 9	SO. BELL CK 3410	491.68	
		9	02/21/84 9	931-9358 - CHECK 228	20.99	
		9	02/24/84 9	SOUTHERN BELL CHECK 2	384.93	
		9	03/20/84 9	SO BELL CK 380	20.99	
		9	03/26/84 9	SO BELL CK 395	281.92	
		9	04/23/84 9	SO BELL CK 493 931-93	20.99	
		9	05/02/84 9	SO BELL CK 505	374.95	
		9	06/04/84 9	932-1128 CK 613	494.39	
		9	06/04/84 9	931-9358 CK 613	20.99	
		9	06/21/84 9	931-9358 CK 3605	20.99	
		9	07/16/84 9	SO BELL CK 3700	450.90	
		9	07/18/84 9	931-9358 CK 3722	20.99	
		9	07/18/84 9	SO. TEL & SEC CK 3717	1,790.25	
		9	08/09/84 9	SO BELL CK 769	193.37	
		9	08/16/84 9	931-9358 CK 882	20.99	
		9	08/29/84 9	932-1128 CK 3840	208.64	

```
RUN DATE: 25-JUL-88                    JOB COST CUMULATIVE TRANSACTION REPORT                           PAGE   71
JOB NAME :

COST-CODE   DESCRIPTION                 TYPE  DATE      REFERENCE                    $ AMOUNT   VOL  AMOUNT

                                          9  08/30/84  931-9358 REFUND                   2.80-
                                          9  10/03/84  932-1128 CK 1009                113.57
                                          9  11/02/84  932-1128 CK 1098                247.30
                                          9  12/05/84  932-1128 CK 4086                245.86
                                          9  12/27/84  932-1128 CK 1321                272.24
                                          9  12/29/84  SO BELL - PAY PHONE              20.99
                                          9  02/08/85  SO BELL CK 1412                 413.93
                                          9  03/12/85  SO BELL CK 1547                 282.76
                                          9  03/28/85  932-1128 - CK 1569              253.39
                                          9  05/02/85  SO BELL CK 4386                 244.00
                                          9  05/23/85  SO BELL CK 4450                 163.55
                                          9  07/08/85  SO BELL CK 4570                 362.22
                                          9  08/12/85  SO BELL CK 4626                 259.17
                                          9  09/10/85  SO BELL CK 1768                 282.17
                                          9  10/16/85  932-1128 CK 4741 FINA            20.51
                                                        -----------------------------------------
           42 ENTRIES FOR THIS TYPE                     SUB-TOTALS                  9,887.25              0.0

            0 ENTRIES FOR THIS TYPE                     SUB-TOTALS                      0.00              0.0
           43 ENTRIES FOR THIS COST CODE                TOTALS                      9,888.30              0.0
010-105    GEN COND - WATER & ELEC - UTIL  ****************************************************************
                                          1  05/22/85  1 P/C CK 4448 REF. DEPO        200.00
                                          1  09/12/85  1 FPL CK 31271 REFUND D        200.00-
                                                        -----------------------------------------
            2 ENTRIES FOR THIS TYPE                     SUB-TOTALS                      0.00              0.0
                                          9  08/02/83  WATER                             0.00
                                          9  09/16/83  WATER                            11.03
                                          9  10/10/83  WATER                            27.08
                                          9  11/10/83  ROSE SEPTIC TANK                 31.04
                                          9  11/11/83  FLORIDA POWER & LIGHT           300.00
                                          9  11/11/83  N.M.B. WATER                    828.82
                                          9  11/30/83  FLORIDA POWER & LIGHT            39.76
                                          9  12/07/83  CITY OF N.M.B. - WATE           732.75
                                          9  12/29/83  F.P.L. CK 3538                   47.68
                                          9  01/26/84  F.P.L. CK 3402                  766.21
                                          9  01/26/84  WATER - CK 3405                 943.41
                                          9  02/10/84  REF. DEPOSIT CHECK #1            39.76
                                          9  02/25/84  N.M.B. CITY HALL CHEC         1,006.67
                                          9  03/01/84  FPL CHECK #252                   60.00
                                          9  03/07/84  NMB CITY HALL CK 281          1,077.87
                                          9  03/20/84  FPL CK 381                    1,112.03
                                          9  04/23/84  NMB WATER CK 500              1,053.32
                                          9  04/23/84  FPL CK 495                       73.02
                                          9  05/08/84  NMB WATER - CK 521            1,228.83
                                          9  06/04/84  CITY OF NMB CK 621               74.90
                                          9  06/21/84  FPL CK 3610                      90.05
                                          9  07/13/84  FPL CK 615                    1,069.07
                                          9  07/19/84  N.M.B. WATER CK 3692          1,221.51
                                          9  08/09/84  FPL CK 3729                      98.36
                                          9  08/16/84  NMB WATER CK 775              1,347.02
                                                        FPL CK 883                    104.30
                                                                                    1,450.57
```

RUN DATE: 25-JUL-88   JOB COST CUMULATIVE TRANSACTION REPORT   PAGE 72
JOB NAME :

COST-CODE	DESCRIPTION	TYPE	DATE	REFERENCE	$ AMOUNT	VOL AMOUNT
		9	09/15/84	FPL CK 995	1,698.11	
		9	09/15/84	NMB - CK 998	69.06	
		9	10/03/84	WATER - CK 1014	101.93	
		9	11/02/84	FPL CK 1099	1,558.24	
		9	11/16/84	CITY OF NMB CK 4019	88.07	
		9	12/05/84	CITY OF NMB CK 4092	88.86	
		9	12/05/84	FPL CK 4088	1,670.62	
		9	12/20/84	FPL CK 1315	1,401.96	
		9	02/08/85	FPL CK REC'D	426.72	
		9	02/08/85	FPL DEP REFUND	1,520.00-	
		9	04/10/85	NMB CITY HALL CK 1619	14.97	
				SUB-TOTALS	19,433.60	0.0
	37 ENTRIES FOR THIS TYPE					
	0 ENTRIES FOR THIS TYPE			SUB-TOTALS	0.00	0.0
	39 ENTRIES FOR THIS COST CODE			TOTALS	19,433.60	0.0
010-106	GEN COND - ICE & CUPS			*******************************		
		1	07/06/83	PETTY CASH	4.58	
		1	07/11/83	BLAIR EQUIP.	2,199.00	
		1	07/15/83	PETTY CASH	11.19	
		1	08/02/83	PETTY CASH	34.83	
		1	08/04/83	BC BUILDERS	177.50	
		1	09/12/83	NATIONAL TOOL	75.60	
		1	09/12/83	WALTON CONTEXT	60.90	
		1	10/10/83	BC BUILDERS	30.87	
		1	10/10/83	WALTON CONTEXT	184.17	
		1	11/10/83	JAR	81.80	
		1	11/10/83	WALTON-CONTEXT	362.25	
		1	01/10/84	NATIONAL TOOL CK 33	142.80	
		1	02/10/84	NATIONAL CONTEXT - CHEC	152.25	
		1	04/10/84	NATIONAL TOOL CK 470	110.25	
		1	04/10/84	JAR CK 466	81.80	
		1	04/10/84	WALTON CK 472	118.47	
		1	04/16/84	P/C CK 595	100.80	
		1	05/11/84	P/C CK 595	100.80	
		1	06/08/84	M.G. ENT CK 695	173.25	
		1	06/20/84	P/C CK 3599	8.12	
		1	07/06/84	P/C CK 3663	24.05	
		1	07/10/84	M.G. ENT CK 764	173.25	
		1	08/10/84	NAT'L CK 838	35.70	
		1	08/10/84	MG ENT CK 849	173.25	
		1	09/10/84	MG ENT CK 926	168.00	
		1	11/09/84	NAT'L TOOL CK 1144	35.70	
		1	04/08/85	M.G. ENT CK 1603	67.20	
		1	05/22/85	P/C CK 4443	24.76	
		1	06/04/85	P/C CK 4468	20.90	
		1	06/28/85	P/C CK 4549	50.40	
		1	07/03/85	P/C CK 4561	10.66	
		1	07/17/85	P/C CK 4584	6.78	
		1	08/16/85	P/C SHEET 8/16/85	1.09	
		1	09/06/85	P/C SLIP 9/6/85	2.28	

RUN DATE: 25-JUL-88
JOB NAME :

JOB COST CUMULATIVE TRANSACTION REPORT

PAGE 73

COST-CODE	DESCRIPTION	TYPE	DATE	REFERENCE	$ AMOUNT	VOL AMOUNT
		1	10/04/85	1 P/C CK 4723	13.31	
				SUB-TOTALS	5,018.56	0.0
		9	11/09/84	BLAIR EQUIP CK 1136	106.58	
		9	12/10/84	BLAIR CK 1235	60.00	
		9	02/08/85	BLAIR EQUIP CK 1458	136.50	
		9	03/10/85	BLAIR EQUIP CK 1519	70.00	
		9	07/10/85	BLAIR EQUIP CK 1716	191.63	
				SUB-TOTALS	564.71	0.0
	35 ENTRIES FOR THIS TYPE			SUB-TOTALS	0.00	0.0
	5 ENTRIES FOR THIS TYPE			TOTALS	5,583.27	0.0
	0 ENTRIES FOR THIS TYPE					
010-107	40 ENTRIES FOR THIS COST CODE					
	GEN COND - TEMP TOILETS					
		1	01/16/85	1 ABLE CK 1392	80.04	
				SUB-TOTALS		
	1 ENTRY FOR THIS TYPE					
		9	08/02/83	ABLE BUILDERS	80.04	0.0
		9	09/01/83	ABLE BUILDERS	90.00	
		9	10/05/83	ABLE BLDRS	49.00	
		9	10/10/83	ABLE BLDRS	135.72	
		9	11/03/83	ABLE BUILDERS & SANIT	49.00	
		9	11/30/83	ABLE BUILDERS	147.00	
		9	12/29/83	ABLE BLDRS CK 3542	98.00	
		9	02/01/84	ABLE BLDRS - CHECK	147.00	
		9	02/15/84	ABLE BLDRS - CHECK #2	180.00	
		9	03/01/84	ABLE BUILDERS CHECK #	386.00	
		9	03/07/84	ABLE BLDRS - CK 265	114.64	
		9	03/20/84	ABLE BLDRS - CK 374	330.00	
		9	03/26/84	ABLE BLDRS CK 393	196.00	
		9	04/05/84	ABLE BLDRS CK 3447	90.24	
		9	04/23/84	ABLE CK 490	92.40	
		9	06/04/84	ABLE BLDRS CK 609	384.00	
		9	06/21/84	ABLE CK 3611	98.44	
		9	09/15/84	ABLE CK 990	304.00	
		9	10/03/84	ABLE BLDRS CK 1007	658.36	
		9	11/02/84	ABLE CK 1095	389.56	
		9	12/27/84	ABLE CK 1319	215.00	
		9	02/08/85	ABLE CK 1409	108.00	
		9	03/12/85	ABLE BLDRS CK 1543	108.00	
				SUB-TOTALS	4,518.36	0.0
	23 ENTRIES FOR THIS TYPE			SUB-TOTALS	0.00	0.0
	0 ENTRIES FOR THIS TYPE			TOTALS	4,598.40	0.0
010-108	24 ENTRIES FOR THIS COST CODE					
	GEN COND- BARRICADES					
		1	08/04/83	1 BC BUILDERS	15.42	
		1	11/10/83	1 MIRON	211.70	
		1	11/10/83	1 AUTOMATIC FASTENERS	84.00	
		1	11/10/83	1 JOHN ABELL CORP.	105.53	
		1	12/09/83	1 HILTI INC.	663.60	

RUN DATE: 25-JUL-88　　　　　　　　JOB COST CUMULATIVE TRANSACTION REPORT　　　　　　　　PAGE 74
JOB NAME :

COST-CODE	DESCRIPTION	TYPE	DATE	REFERENCE	$ AMOUNT	VOL AMOUNT
		1	12/09/83	R.L. LAPP	1,000.00	
		1	12/09/83	MIRON BUILDING PRODS	344.88	
		1	01/10/84	BC BLDRS HDWE - CK 33	158.46	
		1	01/10/84	JOHN ABELL CORP CK 33	492.45	
		1	01/10/84	MIRON BLDG PROD CK 3	1,263.38	
		1	02/10/84	JOHN ABELL CORP. - CH	175.88	
		1	02/10/84	MIRON - CHECK #190	3,039.41	
		1	03/10/84	HILTI - CK 293	407.13	
		1	03/10/84	JOHN ABELL CORP - CK	422.10	
		1	03/10/84	MIRON CK 304	1,554.57	
		1	04/10/84	HILTI CK 465	748.81	
		1	04/10/84	MIRON CK 474	622.78	
		1	05/10/84	HILTI CK 543	266.97	
		1	05/10/84	JOHN ABEL CK 545	105.53	
		1	05/10/84	MIRON CK 551	784.61	
		1	06/08/84	MIRON CK 690	749.40	
		1	07/10/84	MIRON CK 759	131.72	
		1	10/15/84	JAR CK 1045	28.67	
		1	12/10/84	MIRON CK 1246	57.68	
		1	02/26/85	P/C CK 4226	26.69	
		1	03/12/85	P/C CK 1556	2.94	
		1	04/08/85	STODDARD CK 1604	11.34	
		1	09/06/85	P/C SLIP 9/6/85	78.75	
				SUB-TOTALS	13,524.40	0.0
		9	07/14/83	AARON FENCE	4,000.00	
		9	09/20/83	AARON FENCE - FINAL	1,070.00	
		9	02/13/84	P/C CHECK #198	18.50	
				SUB-TOTALS	5,088.50	0.0
	28 ENTRIES FOR THIS TYPE					
	3 ENTRIES FOR THIS TYPE					
	0 ENTRIES FOR THIS TYPE					
	31 ENTRIES FOR THIS COST CODE			TOTALS	18,612.90	0.0

010-109　GEN COND - TRASH CHUTES

		1	10/10/83	CLIFFORD BRYAN	4,500.00	
		1	12/09/83	MIRON BUILDING PRODS	226.36	
		1	01/10/84	AUTOMATIC FASTENERS -	54.16	
		1	03/10/84	MIRON - CK 304	96.73	
		1	04/10/84	HILTI CK 465	67.67	
		1	05/23/84	ROSE SEPTIC TANK CK 3	73.08	
		1	06/08/84	MIRON CK 690	86.59	
		1	06/10/84	MIRON CK 841	47.58	
				SUB-TOTALS	5,152.57	0.0
		6	02/24/84	B/C - 2/21/84 NAPOLEO	337.26-	
		6	02/24/84	B/C 2/17/84 - NAPOLEO	63.24-	
				SUB-TOTALS	400.50-	0.0
	8 ENTRIES FOR THIS TYPE					
	2 ENTRIES FOR THIS TYPE					
	0 ENTRIES FOR THIS TYPE					
	10 ENTRIES FOR THIS COST CODE			TOTALS	4,752.07	0.0

259

```
RUN DATE: 25-JUL-88              JOB COST CUMULATIVE TRANSACTION REPORT                        PAGE  75
JOB NAME :

COST-CODE  DESCRIPTION                       TYPE  DATE     REFERENCE                 $ AMOUNT      VOL  AMOUNT
************************************************************************************************************
010-110    GEN COND - CLNUP DURING CONST       1   02/08/85  1 P/C CK 1419              23.10
                                               1   03/12/85  1 P/C CK 1556               2.58
                                               1   05/20/85  1 P/C CK 4431              35.50
                                                                                  ------------
           3 ENTRIES FOR THIS TYPE                           SUB-TOTALS              61.18           0.0
                                               3   02/08/85  3 CENT FL CK 1460         48.00

           1 ENTRY FOR THIS TYPE                             SUB-TOTALS              48.00           0.0
                                               4   02/24/84  4 B/C 2/20/84 - SAM BLD  137.28-
                                               4   02/24/84  4 B/C 2/20/84 - LEVITZ   137.28-
                                               4   02/24/84  4 B/C 2/20/84 - A&M FOR  137.28-
                                                                                  ------------
           3 ENTRIES FOR THIS TYPE                           SUB-TOTALS             411.84-          0.0
           0 ENTRIES FOR THIS TYPE                           SUB-TOTALS               0.00           0.0
           7 ENTRIES FOR THIS COST CODE                      TOTALS                 302.66-          0.0
010-111    GEN COND - TRASH HAULING            9   12/09/83  9 LATTIMORE HAULING     1,250.00
                                               9   01/10/84  9 DEPENDABLE TRASH CK 3   910.00
                                               9   01/10/84  9 LATTIMORE HAULING CK    780.00
                                               9   02/10/84  9 LATTIMORE - CHECK #18 1,300.00
                                               9   03/10/84  9 LATTIMORE - CK 297    1,490.00
                                               9   04/10/84  9 LATTIMORE CK 468      4,290.00
                                               9   05/10/84  9 LATTIMORE CK 546      5,960.00
                                               9   07/10/84  9 LATTIMORE CK 756      5,370.00
                                               9   08/10/84  9 CENT FL CK 831          557.00
                                               9   08/10/84  9 LATTIMORE CK 836      7,535.00
                                               9   09/07/84  9 LATTIMORE CK 3849     3,880.00
                                               9   10/15/84  9 LATTIMORE CK 1046     2,290.00
                                               9   11/09/84  9 LATTIMORE CK 1142     4,500.00
                                               9   12/10/84  9 LATTIMORE CK 1242     5,985.00
                                               9   01/10/85  9 LATTIMORE CK 1371     3,875.00
                                               9   02/08/85  9 LATTIMORE CK 1463     2,775.00
                                               9   03/10/85  9 LATTIMORE CK 1524     1,085.00
                                               9   04/04/85  9 LATTIMORE CK 4305     1,085.00
                                               9   05/03/85  9 LATTIMORE CK 4392       465.00
                                               9   06/10/85  9 LATTIMORE CK 1680     1,705.00
                                               9   07/10/85  9 LATTIMORE CK 1717     1,240.00
                                                                                  ------------
          21 ENTRIES FOR THIS TYPE                           SUB-TOTALS          57,827.00           0.0
           0 ENTRIES FOR THIS TYPE                           SUB-TOTALS               0.00           0.0
          21 ENTRIES FOR THIS COST CODE                      TOTALS              57,827.00           0.0
010-112    GEN COND - TRASH CONTAINERS         1   02/08/85  1 NAT'L TOOL CK 1465       3.15

           1 ENTRY FOR THIS TYPE                             SUB-TOTALS               3.15
                                               9   08/15/83  9 UNITED SANITATION      115.60
                                               9   09/16/83  9 UNITED SANITATION      346.80
                                               9   10/10/83  9 UNITED SANITATION      265.20
```

RUN DATE: 25-JUL-88   JOB COST CUMULATIVE TRANSACTION REPORT   PAGE 76
JOB NAME :

COST-CODE	DESCRIPTION	TYPE	DATE	REFERENCE	$ AMOUNT	VOL AMOUNT
		9	11/11/83	9 UNITED SANITATION	142.80	
		9	12/29/83	9 UNITED SANITATION	142.80	
		9	01/26/84	9 UNITED SANITATION C	285.60	
		9	03/07/84	9 UNITED SANITATION CK	142.80	
		9	04/23/84	9 UNITED SAN CK 494	20.40	
		9	05/08/84	9 UNITED SANITATION CK	265.20	
		9	06/21/84	9 UNITED - CK 3603	632.40	
		9	07/15/84	9 UNITED SAN CK 3688	387.60	
		9	07/16/84	9 ABLE BLDRS CK 3697	500.44	
		9	08/09/84	9 ABLE BLDRS CK 767	707.24	
		9	08/09/84	9 UNITED SANIT CK 770	265.20	
		9	09/15/84	9 UNITED SAN CK 994	142.80	
		9	10/05/84	9 UNITED SAN CK 3326	142.80	
		9	11/16/84	9 UNITED SAN. CK 4026	20.40	
		9	12/10/84	9 UNITED SAN CK 1288	1,218.90	
		9	01/16/85	9 UNITED SAN CK 1395	1,575.90	
		9	03/12/85	9 UNITED SANITATION CK	999.60	
		9	04/10/85	9 UNITED SAN CK 1613	493.00	
		9	05/09/85	9 UNITED SAN CK 1660		
				SUB-TOTALS	9,078.68	0.0
				SUB-TOTALS	0.00	0.0
				TOTALS	9,081.83	0.0

 22 ENTRIES FOR THIS TYPE
  0 ENTRIES FOR THIS TYPE
 23 ENTRIES FOR THIS COST CODE

010-113 GEN COND - PUNCH LIST

		1	04/08/85	T-SQUARE CK 1595	14.70	
		1	08/16/85	P.C SHEET 8/16/85	224.97	
		1	09/10/85	B.C. CK 1766	339.24	
		1	10/04/85	P/C CK 4723	18.62	
		1	10/10/85	BC CK 4732	284.32	
		1	10/10/85	SAF-T-GREEN CK 4728	510.81	
		1	12/31/85	P/C SLIP 12/1	49.43	
				SUB-TOTALS	1,442.09	
				TOTALS	1,442.09	

 7 ENTRIES FOR THIS TYPE
 7 ENTRIES FOR THIS COST CODE

010-116 GEN COND - TOOLS & EQUIPMENT

		1	07/07/83	PETTY CASH	70.00	
		1	08/04/83	TURNBERRY CORP	1,932.00	
		1	08/30/83	PETTY CASH	47.06	
		1	09/09/83	HA TOOL & SAW	7.50	
		1	09/12/83	FLORIDA LEVEL	105.00	
		1	09/12/83	JAR SUPPLIES	79.43	
		1	09/12/83	NATIONAL TOOL	186.32	
		1	09/12/83	PHN EQUIP	67.80	
		1	09/13/83	PETTY CASH	23.05	
		1	09/16/83	LAPP, R.L.	1,450.00	
		1	09/22/83	PETTY CASH	92.11	
		1	10/10/83	PHN EQUIP	185.40	
		1	10/10/83	AAA TOOL	25.20	
		1	10/10/83	AUTOMATIC FASTENERS	732.65	
		1	10/10/83	BLANCHARD	45.99	

261

RUN DATE: 25-JUL-88 JOB COST CUMULATIVE TRANSACTION REPORT PAGE 80
JOB NAME :

COST-CODE	DESCRIPTION	TYPE	DATE	REFERENCE	$ AMOUNT	VOL AMOUNT
	21 ENTRIES FOR THIS TYPE	9	02/10/84	SUB-TOTALS	11,980.67	0.0
		9	07/10/85	BLANCHARD - CHECK #17	522.80	
				A-1 RENT ALLS CK 1721	119.70	
	2 ENTRIES FOR THIS TYPE			SUB-TOTALS	642.50	0.0
	0 ENTRIES FOR THIS TYPE			SUB-TOTALS	0.00	0.0
	168 ENTRIES FOR THIS COST CODE			TOTALS	31,202.69	0.0
010-117	GEN COND - TOPPING OUT PARTY	1	04/23/84	P/C CK 501	195.56	
	1 ENTRY FOR THIS TYPE			SUB-TOTALS	195.56	0.0
	0 ENTRIES FOR THIS TYPE			SUB-TOTALS	0.00	0.0
	1 ENTRY FOR THIS COST CODE			TOTALS	195.56	0.0
010-119	GEN COND - SCAFFOLDING	9	05/31/84	P/C CK 3555	110.00	
		9	08/10/84	ANTHES CK 850	115.50	
		9	09/10/84	ANTHES CK 927	57.75	
		9	10/10/84	ANTHES CK 1055	57.75	
		9	11/09/84	ANTHES CK 1153	57.75	
		9	03/10/85	PATENT SCAF. CK 1541	168.60	
		9	04/08/85	PATENT SCAFFOLD CK 16	81.45	
		9	05/09/85	PATENT SCAF CK 1671	54.93	
		9	06/10/85	PATENT SCAF. CK 1692	573.89	
	9 ENTRIES FOR THIS TYPE			SUB-TOTALS	1,277.02	0.0
	0 ENTRIES FOR THIS TYPE			SUB-TOTALS	0.00	0.0
	9 ENTRIES FOR THIS COST CODE			TOTALS	1,277.02	0.0
010-120	GEN COND - TRUCK EXPENSE	1	06/24/83	PETTY CASH	70.00	
		1	06/29/83	PETTY CASH	36.50	
		1	07/06/83	PETTY CASH	34.00	
		1	07/15/83	PETTY CASH	63.49	
		1	08/02/83	PETTY CASH	109.00	
		1	08/10/83	PETTY CASH	40.00	
		1	08/17/83	PETTY CASH	113.33	
		1	08/30/83	PETTY CASH	120.50	
		1	09/13/83	PETTY CASH	40.00	
		1	09/22/83	PETTY CASH	60.00	
		1	10/07/83	PETTY CASH	58.00	
		1	10/13/83	PETTY CASH	15.00	
		1	10/13/83	PETTY CASH	20.00	
		1	11/03/83	PETTY CASH	24.00	
		1	11/16/83	PETTY CASH	123.59	
		1	11/16/83	PETTY CASH	91.34	
		1	11/22/83	PETTY CASH	19.95	
		1	11/22/83	PETTY CASH	25.00	
		1	12/16/83	PETTY CASH	79.74	

RUN DATE: 25-JUL-88   JOB COST CUMULATIVE TRANSACTION REPORT   PAGE 81

JOB NAME :

COST-CODE	DESCRIPTION	TYPE	DATE	REFERENCE	$ AMOUNT	VOL AMOUNT
		1	12/16/83	PETTY CASH	25.61	
		1	12/23/83	PETTY CASH	24.82	
		1	12/29/83	PETTY CASH Z.P. CK	84.04	
		1	01/06/84	PETTY CASH Z.P. C	50.85	
		1	01/16/84	PETTY CASH - CK #3359	352.47	
		1	01/19/84	PETTY CASH D.A. CK	89.25	
		1	02/01/84	PETTY CASH - CHECK #1	115.87	
		1	02/06/84	PETTY CASH - CHECK #1	39.00	
		1	02/13/84	P/C CHECK #198	42.73	
		1	02/15/84	PETTY CASH - CHECK #2	19.00	
		1	02/21/84	HIGHLAND AUTOMOTIVE C	149.89	
		1	03/01/84	PETTY CASH - CHECK #2	104.95	
		1	03/20/84	HIGHLAND AUTO CK 3449	42.00	
		1	03/20/84	P/C CK 385	15.00	
		1	03/20/84	P/C CK 385	2.00	
		1	03/26/84	P/C CK 397	34.00	
		1	03/26/84	P/C CK 397	4.00	
		1	03/29/84	P/C CK 410	36.00	
		1	04/05/84	P/C CK 3462	5.25	
		1	04/05/84	P/C CK 3462	13.00	
		1	04/13/84	P/C CK 489	13.00	
		1	04/13/84	P/C CK 489	21.00	
		1	04/23/84	P/C CK 501	14.00	
		1	04/23/84	P/C CK 501	19.25	
		1	05/02/84	P/C CK 511	64.00	
		1	05/08/84	PETTY CASH CK 522	39.35	
		1	05/11/84	P/C CK 595	17.00	
		1	05/11/84	P/C CK 595	10.00	
		1	05/23/84	P/C CK 3542	51.78	
		1	05/31/84	P/C CK 3555	23.00	
		1	05/31/84	P/C CK 3555	15.00	
		1	06/04/84	HIGHLAND AUTO CK 3567	352.38	
		1	06/04/84	P/C CK 622	33.60	
		1	06/21/84	P/C CK 3599	48.00	
		1	06/28/84	P/C CK 3629	82.45	
		1	06/28/84	P/C CK 3629	15.00	
		1	07/06/84	P/C CK 3663	54.00	
		1	07/06/84	P/C CK 3663	19.50	
		1	07/10/84	HIGHLAND AUTO CK 3668	150.15	
		1	07/18/84	P/C CK 3726	44.00	
		1	07/19/84	P/C CK 3726	27.24	
		1	07/24/84	P/C CK 3741	24.86	
		1	07/24/84	P/C CK 3741	55.00	
		1	07/30/84	P/C CK 3752	20.40	
		1	07/30/84	P/C CK 3752	13.00	
		1	08/01/84	P/C CK 3759	24.00	
		1	08/01/84	P/C CK 3759	15.80	
		1	08/09/84	P/C CK 776	49.07	
		1	08/09/84	P/C CK 776	15.20	
		1	08/09/84	P/C CK 776	17.60	
		1	08/15/84	P/C CK 888	75.00	

```
RUN DATE: 25-JUL-88                    JOB COST CUMULATIVE TRANSACTION REPORT                          PAGE 83
JOB NAME :

COST-CODE   DESCRIPTION                 TYPE  DATE     REFERENCE                    $ AMOUNT      VOL AMOUNT

                                         1  05/20/85 1  P/C CK 4431                    83.87
                                         1  05/20/85 1  P/C CK 4431                    40.83
                                         1  05/22/85 1  P/C CK 4443                    56.00
                                         1  06/04/85 1  P/C CK 4468                     2.66
                                         1  06/04/85 1  P/C CK 4468                    89.64
                                         1  06/11/85 1  P/C CK 4503                    43.50
                                         1  06/11/85 1  P/C CK 4403                   336.02
                                         1  06/18/85 1  P/C CK 4522                    60.66
                                         1  06/28/85 1  P/C CK 4549                    77.44
                                         1  07/03/85 1  P/C CK 4561                   101.60
                                         1  07/17/85 1  P/C CK 4584                    19.14
                                         1  07/17/85 1  P/C CK 4584                    48.19
                                         1  07/23/85 1  P/C CK 4591                    55.75
                                         1  07/26/85 1  P/C SHEET 7/26/85              64.63
                                         1  08/16/85 1  P/C SHEET 8/16/85             226.55
                                         1  09/06/85 1  GENE HOLMAN CK 4660           219.36
                                         1  09/06/85 1  P/C SLIP 9/6/85               316.18
                                         1  09/10/85 1  HIGHLAND AUTO CK 1770          76.91
                                         1  10/04/85 1  P/C CK 4723                   170.16
                                         1  12/31/85 1  P/C TICKET 12/1               130.23
                                         1  12/31/85 1  P/C TICKET 12/1                30.00
                                                                                  -----------
                                                        SUB-TOTALS                  9,316.85

                                         3  04/08/85 3  HIGHLAND AUTO CK 1597       1,109.94
                                         3  04/10/85 3  HIGHLAND CK 1614               63.00
                                                                                  -----------
                                                        SUB-TOTALS                    172.94

  142 ENTRIES FOR THIS TYPE                                                                          0.0

    2 ENTRIES FOR THIS TYPE                                                                          0.0
                                                        SUB-TOTALS                      0.00         0.0
                                                        TOTALS                      9,489.79         0.0
                                        *********************************************************************
010-121   GEN COND - RUBBING & PATCHING
                                         1  09/12/83 1  BC BUILDERS                    92.92
                                         1  10/10/83 1  BC BUILDERS                     6.11
                                         1  10/10/83 1  WALTON CONTEXT                 26.15
                                         1  11/10/83 1  WALTON CONTEXT                 83.21
                                         1  01/10/84 1  WALTON CONTEXT       CK 3      59.43
                                         1  02/10/84 1  SUNSHINE CONT. - CHEC          43.47
                                         1  05/10/84 1  SUNSHINE SUP CK 554            48.30
                                         1  08/10/84 1  MIRON CK 841                   65.18
                                         1  02/08/85 1  M.G. ENT CK 1469               31.50
                                         1  03/10/85 1  B.C. BLDRS CK 1520             21.48
                                         1  03/28/85 1  P/C CK 1574                    17.33
                                         1  06/10/85 1  BC BLDRS CK 1679                4.04
                                         1  06/18/85 1  P/C CK 4522                    65.52
                                         1  06/28/85 1  P/C CK 4549                    12.50
                                         1  07/03/85 1  P/C CK 4561                   111.30
                                         1  07/17/85 1  P/C CK 4584                     4.38
                                         1  07/23/85 1  P/C CK 4591                    15.70
                                         1  08/16/85 1  P/C SHEET 8/16/85             200.00
                                         1  09/10/85 1  B.C. CK 1766                   70.93
                                         1  09/10/85 1  SUNSHINE CONT CK 1771          81.90
```

RUN DATE: 25-JUL-88
JOB NAME :

JOB COST CUMULATIVE TRANSACTION REPORT

PAGE 84

COST-CODE	DESCRIPTION	TYPE	DATE	REFERENCE	$ AMOUNT	VOL AMOUNT
	20 ENTRIES FOR THIS TYPE			SUB-TOTALS	1,098.35	0.0
		9	06/10/85	9 ED. GREEN CK 1676	205.89	
				SUB-TOTALS	205.89	0.0
				TOTALS	1,304.24	0.0
010-122	1 ENTRY FOR THIS TYPE					
	21 ENTRIES FOR THIS COST CODE					
	GEN COND - SURVEYS & TESTING					
		9	07/10/85	9 PETER DELA ROSA CK 17	225.00	
				SUB-TOTALS	225.00	0.0
	1 ENTRY FOR THIS TYPE					
				SUB-TOTALS	0.00	0.0
				TOTALS	225.00	0.0
010-123	0 ENTRIES FOR THIS TYPE					
	1 ENTRY FOR THIS COST CODE					
	GEN COND - PROJECT SIGNS					
		1	08/04/83	1 PAN AMER LUMBER	84.00	
		1	08/10/84	1 MIRON CK 841	2.75	
	2 ENTRIES FOR THIS TYPE			SUB-TOTALS	86.75	
		9	07/20/83	9 JIM HALL	945.00	
		9	08/02/83	9 PETTY CASH	200.00	
	2 ENTRIES FOR THIS TYPE			SUB-TOTALS	1,145.00	0.0
				TOTALS	1,231.75	0.0
	0 ENTRIES FOR THIS TYPE					
	4 ENTRIES FOR THIS COST CODE					
010-124	GEN COND - RENTAL EQUIPMENT					
		1	06/04/84	1 P/C CK 622	20.00	
		1	11/02/84	1 P/C CK 1110	34.20	
		1	05/08/85	1 P/C CK 4402	6.06	
		1	05/08/85	1 P/C CK 4402	5.00	
	4 ENTRIES FOR THIS TYPE			SUB-TOTALS	65.26	0.0
		3	09/12/83	3 BLANCHARD	41.06	
		3	09/12/83	3 CENTRAL FLA	360.00	
		3	09/12/83	3 LATTIMORE	130.00	
		3	09/16/83	3 LAPP, R.L.	150.00	
		3	10/07/83	3 PETTY CASH	11.50	
		3	10/10/83	3 CENTRAL FLA EQUIP	702.00	
		3	11/10/83	3 BOYS PLUMBING	217.75	
		3	11/10/83	3 R.L. LAPP	100.00	
		3	11/22/83	3 PETTY CASH	300.00	
		3	12/09/83	3 CENTRAL FLA EQUIPMENT	432.00	
		3	01/10/84	3 CENTRAL FLA EQUIP - C	1,206.00	
		3	02/10/84	3 CENTRAL FLA - CHECK #	576.00	
		3	03/07/84	3 PETTY CASH - CK 267	40.00	
		3	03/10/84	3 CENTRAL FLA - CK 291	356.00	
		3	04/10/84	3 CENTRAL FLA CK 463	576.00	
		3	05/10/84	3 CENTRAL EQUIP CK 542	414.00	
		3	06/08/84	3 CENTRAL FLA CK 680	304.00	
		3	07/10/84	3 CENTRAL FLA CK 751	285.00	
		3	08/10/84	3 BLANCHARD CK 829	186.90	

RUN DATE: 25-JUL-88   JOB COST CUMULATIVE TRANSACTION REPORT   PAGE 85

JOB NAME :

COST-CODE	DESCRIPTION	TYPE	DATE	REFERENCE	$ AMOUNT	VOL AMOUNT
		3	08/10/84	3 CENT FL CK 831	266.00	
		3	09/10/84	3 BLANCHARD CK 911	300.46	
		3	09/10/84	3 CENTRAL FLA CK 913	342.00	
		3	11/09/84	3 SNEAD CANIPE CK 1146	441.00	
		3	12/10/84	3 CENTRAL FLA CK 1239	125.00	
		3	01/10/85	3 BLANCHARD CK 1367	159.08	
		3	02/08/85	3 BLANCHARD CK 1459	143.17	
		3	02/08/85	3 METRO TRUCKING CK 147	176.00	
		3	03/28/85	3 BERES WAITE CK 1577 P	225.00	
		3	05/09/85	3 SNEAD-CANIPE CK 1658	1,032.00	
		3	06/10/85	3 HARRIS BACKHOE CK 169	74.00	
		3	08/08/85	3 HARRIS BACKHOE CK 176	74.00	
	31 ENTRIES FOR THIS TYPE			SUB-TOTALS	9,745.92	0.0
		9	01/10/84	9 PHN EQUIP CK 3327	189.00	
		9	02/10/84	9 R.L. LAPP - CHECK #19	900.00	
	2 ENTRIES FOR THIS TYPE			SUB-TOTALS	1,089.00	0.0
	0 ENTRIES FOR THIS TYPE				0.00	0.0
010-127	37 ENTRIES FOR THIS COST CODE			TOTALS	10,900.18	0.0
	RUNNER					
		1	04/10/85	1 P/C CK 1620	36.00	
		1	04/22/85	1 P/C CK 4349	33.45	
	2 ENTRIES FOR THIS TYPE			SUB-TOTALS	69.45	0.0
010-131	2 ENTRIES FOR THIS COST CODE			TOTALS	69.45	0.0
	INCLEMENT WEATHER					
		1	11/09/84	1 MIRON CK 1147	623.57	
	1 ENTRY FOR THIS TYPE			SUB-TOTALS	623.57	
010-132	1 ENTRY FOR THIS COST CODE			TOTALS	623.57	0.0
	CHRISTMAS PARTY					
		1	12/21/84	1 FOREMOST-CK 4103 PER	177.44	
	1 ENTRY FOR THIS TYPE			SUB-TOTALS	177.44	
	0 ENTRIES FOR THIS TYPE				0.00	0.0
010-133	1 ENTRY FOR THIS COST CODE			TOTALS	177.44	0.0
	MOVE TO WAREHOUSE					
		9	06/08/85	9 P/C CK 4522	175.55	
		9	08/08/85	9 PUBLIC STORAGE CK 176	93.45	
		9	09/10/85	9 PUB STORAGE CK 1776	93.45	
		9	10/10/85	9 PUBLIC STORAGE CK 473	93.45	
	4 ENTRIES FOR THIS TYPE			SUB-TOTALS	455.90	0.0
	0 ENTRIES FOR THIS TYPE				0.00	0.0
011-101	4 ENTRIES FOR THIS COST CODE			TOTALS	455.90	0.0
	HOISTING - HOUSE ELEVATOR					
		1	11/02/84	1 P/C CK 1110	50.00	

RUN DATE: 25-JUL-88   JOB COST CUMULATIVE TRANSACTION REPORT   PAGE 86
JOB NAME :

COST-CODE	DESCRIPTION	TYPE	DATE	REFERENCE	$ AMOUNT	VOL AMOUNT
				SUB-TOTALS	50.00	0.0
				TOTALS	50.00	0.0
011-102	HOISTING - MATERIAL HOIST					
	1 ENTRY FOR THIS TYPE	1	06/28/84	P/C CK 3629	32.71	
				SUB-TOTALS	32.71	0.0
		3	01/10/84	CENTRAL FLA EQUIP - C	216.00	
		3	01/13/84	FLA CRANE - CK #3357	11,050.87	
		3	02/10/84	FLA CRANE - CHECK #16	5,717.25	
		3	03/10/84	FLA CRANE - CK 284	11,337.21	
		3	04/10/84	FLA CRANE CK 460	6,548.85	
		3	04/17/84	FLA CRANE CK 3477	497.25	
		3	05/10/84	FLA CRANE CK 528	8,528.85	
		3	06/08/84	FL CRANE CK 677	8,548.86	
		3	12/10/84	NAPOLEON CK 1259	3,200.00	
	9 ENTRIES FOR THIS TYPE					
		9	08/06/84	ELITE GUARD CK 3786 P	55,645.14	
		9	08/30/84	ELITE GUARD CK 3829 P	1,579.50	
		9	09/07/84	ELITE GUARD CK 3850 P	812.50	
	3 ENTRIES FOR THIS TYPE				2,384.50	
				SUB-TOTALS	4,776.50	0.0
	0 ENTRIES FOR THIS TYPE			TOTALS	60,454.35	0.0
	13 ENTRIES FOR THIS COST CODE					
011-103	HOISTING - CRANE					
		3	08/04/83	FLORIDA CRANE	735.00	
		3	08/04/83	NAPOLEON STEEL	3,066.29	
		3	09/12/83	FLORIDA CRANE	14,599.56	
		3	09/12/83	NAPOLEON STEEL	5,522.66	
		3	10/10/83	CENTRAL FLA EQUIP	72.00	
		3	10/10/83	FLORIDA CRANE	11,061.31	
		3	10/10/83	NAPOLEON STEEL	2,279.50	
		3	11/10/83	FLORIDA CRANE	60,159.17	
		3	12/09/83	CENTRAL FLA EQUIPMENT	140.00	
		3	12/09/83	FLORIDA CRANE - CK #3	45,131.17	
		3	01/10/84	NAPOLEON STEEL CK 3	5,163.25	
		3	01/13/84	FLA CRANE - CK #3357	26,913.74	
		3	02/10/84	FLA CRANE - CHECK #16	45,547.20	
		3	02/10/84	NAPOLEON STEEL - CHEC	7,447.40	
		3	02/24/84	CENTRAL FLA - CHECK #	108.00	
		3	03/10/84	MATERIAL HANDLERS CHE	403.75	
		3	03/10/84	FLA CRANE - CK 284	76,546.92	
		3	03/10/84	NAPOLEON STEEL - CK 2	142.00	
		3	04/10/84	ALAN W. SMITH CK 477	120.00	
		3	04/10/84	FLA CRANE CK 460	64,003.52	
		3	04/10/84	MAT HNDLRS CK 476	403.75	
		3	04/17/84	FLA CRANE CK 3477	5,958.50	
		3	05/10/84	CREST CONC. CK 556	877.50	
		3	05/10/84	FLA CRANE CK 528	42,652.37	

RUN DATE: 25-JUL-88   JOB COST CUMULATIVE TRANSACTION REPORT   PAGE 87

JOB NAME :

COST-CODE	DESCRIPTION	TYPE	DATE	REFERENCE	$ AMOUNT	VDL AMOUNT
		3	06/08/84	3 FL CRANE CK 677	45,879.96	
		3	08/10/84	3 A.W.S.I. CK 847	580.00	
		3	09/10/84	3 AWSI CK 923	200.00	
		3	09/10/84	3 MIAMI CRANE CK 924	13,020.00	
		3	10/15/84	3 A.W.S.I. CK 1051	710.00	
		3	10/15/84	3 MIAMI CRANE CK 1052	4,950.77	
		3	10/15/84	3 NAPOLEON CK 1039	264.00	
		3	11/09/84	3 NAPOLEON CK 1159	24.00	
		3	11/09/84	3 MIAMI CRANE CK 1149	297.50	
		3	12/10/84	3 A.W.S.I. CK 1148	2,239.25	
		3	12/10/84	3 A.W.S.I. CK 1248	855.00	
		3	12/10/84	3 CREST CK 1250	247.50	
		3	01/10/85	3 MIAMI CRANE CK 1249	8,023.35	
		3	01/10/85	3 ALAN W. SMITH CK 1375	517.50	
		3	01/10/85	3 MIAMI CRANE CK 1376	9,387.92	
		3	01/10/85	3 CREST CK 1377	660.00	
		3	02/08/85	3 MIAMI CRANE CK 1467	2,640.00	
		3	03/10/85	3 ALAN W. SMITH CK 1531	1,451.25	
		3	03/10/85	3 MIAMI CRANE CK 1532	506.25	
		3	04/08/85	3 A.W.S.I. CK 1599	14,146.49	
		3	04/08/85	3 MIAMI CRANE CK 1600	1,696.25	
		3	05/09/85	3 CREST CK 1601	770.00	
		3	05/09/85	3 HARRIS BACKHOE CK 167	129.50	
		3	05/09/85	3 CREST CONC CK 1667	680.00	
		3	06/10/85	3 A.W.S.I. CK 1666	225.00	
		3	06/10/85	3 BOB MOORE'S CK 1697	200.00	
		3	06/10/85	3 CREST CONC. CK 1686	165.00	
		3	06/10/85	3 MIAMI CRANE CK 1685	402.50	
		3	09/10/85	3 CREST CONC CK 1773	330.00	

53 ENTRIES FOR THIS TYPE   SUB-TOTALS   529,958.55   0.0

0 ENTRIES FOR THIS TYPE   SUB-TOTALS   0.00   0.0
53 ENTRIES FOR THIS COST CODE   TOTALS   529,958.55   0.0
012-108 LAYOUT - MATERIAL & EQUIPMENT
****

		1	08/04/83	1 BC BUILDERS	210.61	
		1	08/04/83	1 PAN AMER LUMBER	81.03	
		1	08/10/83	1 PAN AMER LUMBER	27.76	
		1	08/10/83	1 KEMRON	19.95	
		1	09/12/83	1 BC BUILDERS	172.25	
		1	09/12/83	1 JAR SUPPLIES	18.58	
		1	09/16/83	1 LAPP, R.L.	150.00	
		1	10/10/83	1 WALTON CONTEXT	18.75	
		1	10/10/83	1 B.C. BUILDERS	114.21	
		1	10/18/83	1 JOHN ABELL CORP	107.18	
		1	11/10/83	1 BC BUILDERS	11.50	
		1	11/10/83	1 JAFFES	27.49	
		1	11/16/83	1 PETTY CASH	3.03	
		1	12/09/83	1 BC BUILDERS HARDWARE	221.22	
		1	12/09/83	1 FLORIDA LEVEL & TRANS	3.68	
		1	12/09/83	1 MIRON BUILDING PRODS	33.60	

RUN DATE: 25-JUL-88   JOB COST CUMULATIVE TRANSACTION REPORT   PAGE 88

JOB NAME :

COST-CODE	DESCRIPTION	TYPE	DATE	REFERENCE	$ AMOUNT	VOL AMOUNT
		1	12/16/83	PETTY CASH	29.70	
		1	01/10/84	B.C. BLDRS HDWE - CK	127.87	
		1	01/19/84	PETTY CASH - D.A. CK	332.22	
		1	02/10/84	BC BLDRS - CHECK #175	113.82	
		1	02/10/84	FLA LEVEL - CHECK #17	2,001.93	
		1	02/10/84	GRABBER - CHECK #178	99.34	
		1	02/13/84	P/C CHECK #198	13.86	
		1	03/10/84	BC BLDRS - CK 290	19.04	
		1	03/10/84	GRABBER - CK 292	22.30	
		1	03/10/84	MIRON - CK 304	71.56	
		1	04/10/84	FLA LVL & TRST CK 464	19.95	
		1	04/10/84	JAR CK 466	8.92	
		1	04/10/84	MIRON CK 474	14.69	
		1	05/10/84	BC BLDRS CK 541	90.75	
		1	05/10/84	MIRON CK 551	8.82	
		1	06/08/84	BC BLDRS CK 679	57.12	
		1	06/08/84	FLA LEVEL CK 681	36.05	
		1	07/10/84	MIRON CK 759	37.99	
		1	07/10/84	BC BLDRS CK 750	4.32	
		1	08/10/84	BC CK 830	66.22	
		1	08/10/84	MIRON CK 841	11.87	
		1	09/10/84	GRABBER CK 914	77.45	
		1	10/15/84	GRABBER CK 1043	11.25	
		1	10/15/84	MIRON CK 1048	17.29	
		1	11/09/84	BC BLDRS CK 1138	35.76	
		1	11/09/84	MIRON CK 1147	57.11	
		1	12/13/84	P/C CK 1301	8.35	
		1	12/27/84	P/C CK 1327	10.45	
		1	02/08/85	B.C. BLDRS CK 1476	46.35	
		1	03/10/85	B.C. BLDRS CK 1520	8.96	
		1	03/10/85	FLA LEVEL CK 1521	77.25	
		1	03/28/85	P/C CK 1574	5.83	
		1	04/08/85	B.C. HDWE CK 1587	92.71	
				SUB-TOTALS	4,857.94	0.0
		3	01/10/84	FLA LEVEL CK 3343	35.70	
		3	12/10/84	FLA LEV CK 1240	378.00	
				SUB-TOTALS	413.70	0.0
				SUB-TOTALS	0.00	0.0
				TOTALS	5,271.64	0.0

49 ENTRIES FOR THIS TYPE
2 ENTRIES FOR THIS TYPE
0 ENTRIES FOR THIS TYPE
51 ENTRIES FOR THIS COST CODE

013-101   MILLWORK - S.C. PRE-HUNG

		1	01/10/84	MIRON BLDG PROD CK	33.60	
		1	02/10/84	MIRON - CHECK #190	42.00	
		1	05/10/84	MIRON CK 551	16.59	
		1	03/10/85	SUNSHINE CK 1529	128.63	
		1	06/28/85	P/C CK 4549	36.22	
				SUB-TOTALS	257.04	0.0
				TOTALS	257.04	0.0

5 ENTRIES FOR THIS TYPE
5 ENTRIES FOR THIS COST CODE

RUN DATE: 25-JUL-88
JOB COST CUMULATIVE TRANSACTION REPORT
PAGE 89

JOB NAME :

COST-CODE	DESCRIPTION	TYPE	DATE	REFERENCE	$ AMOUNT	VOL AMOUNT
013-103	MILLWORK - H/M DOORS					
		1	05/22/85	1 P/C CK 4443	18.65	
				SUB-TOTALS	18.65	0.0
				TOTALS	18.65	0.0
013-105	MILLWORK - MEDICINE CABINETS					
	1 ENTRY FOR THIS TYPE					
	1 ENTRY FOR THIS COST CODE	1	12/10/84	1 MIRON CK 1246	41.20	
				SUB-TOTALS	41.20	0.0
				TOTALS	41.20	0.0
014-101	ROUGH CARP - TUBS - FRAMING					
	1 ENTRY FOR THIS TYPE					
	1 ENTRY FOR THIS COST CODE	1	05/10/84	1 MIRON CK 551	810.60	
		1	07/10/84	1 MIRON CK 759	377.44	
		1	08/10/84	1 MIRON CK 841	727.09	
		1	09/10/84	1 MIRON CK 922	267.13	
		1	11/09/84	1 PAN AM LUMB CK 1145	1,913.41	
		1	03/10/85	1 FLA LUMBER CK 1542	48.30	
				SUB-TOTALS	4,143.97	0.0
				TOTALS	4,143.97	0.0
014-102	ROUGH CARP - SUN DECKS					
	6 ENTRIES FOR THIS TYPE					
	6 ENTRIES FOR THIS COST CODE	1	11/09/84	1 PAN AM LUMB CK 1145	1,873.52	
		1	12/10/84	1 BC BLDRS CK 1237	57.12	
		1	12/10/84	1 MIRON CK 1246	41.54	
		1	02/08/85	1 PAN AM LUMB CK 1477	1,011.87	
		1	03/10/85	1 FLA LUMBER CK 1542	131.71	
				SUB-TOTALS	3,415.76	0.0
				TOTALS	3,415.76	0.0
014-103	ROUGH CARP - SHADOW BOXES					
	5 ENTRIES FOR THIS TYPE					
	5 ENTRIES FOR THIS COST CODE	1	11/09/84	1 PAN AM LUMB CK 1145	147.52	
				SUB-TOTALS	147.52	0.0
				TOTALS	147.52	0.0
014-104	ROUGH CARP - PAV & ROOF DECK					
	1 ENTRY FOR THIS TYPE					
	1 ENTRY FOR THIS COST CODE	1	04/10/84	1 MIRON CK 474	623.28	
		1	12/10/84	1 MIRON CK 1246	2,150.76	
		1	01/10/85	1 PAN AM CK 1373	205.02	
		1	03/10/85	1 FLA LUMBER CK 1542	510.60	
		1	05/09/85	1 FLA LUMBER CK 1672	1,643.25	
		1	06/10/85	1 PAN AM LUMBER CK 1682	831.60	
		1	08/08/85	1 FLA LUMBER CK 1761	33.22	
					58.35	
				SUB-TOTALS	6,056.08	0.0
				TOTALS	6,056.08	0.0
014-105	ROUGH CARP-GAZEBO ROOF & DECK					
	8 ENTRIES FOR THIS TYPE					
	8 ENTRIES FOR THIS COST CODE	1	03/10/85	1 FLA LUMBER CK 1542	368.12	
		1	03/12/85	1 P/C CK 1556	35.70	
		1	06/10/85	1 FLA LUMBER CK 1693	422.71	

RUN DATE: 25-JUL-88   JOB COST CUMULATIVE TRANSACTION REPORT   PAGE 90

JOB NAME :

COST-CODE	DESCRIPTION	TYPE	DATE	REFERENCE	$ AMOUNT	VOL AMOUNT
014-106	ROUGH CARP - TRELLIS #1			SUB-TOTALS TOTALS	826.53 826.53	0.0 0.0
		1	02/08/85	PAN AM LUMBER CK 1477	13.47	
		1	03/10/85	PAN AM LMBR CK 1526	190.16	
		1	03/10/85	FLA LUMBER CK 1542	1,718.95	
		1	05/09/85	FLA LUMBER CK 1672	1,127.28	
		1	09/10/85	FLA LUMBER CK 1774	105.95	
	5 ENTRIES FOR THIS TYPE			SUB-TOTALS	3,155.81	0.0
	5 ENTRIES FOR THIS COST CODE			TOTALS	3,155.81	0.0
014-107	ROUGH CARP - TRELLIS #2					
		1	03/10/85	FLA LUMBER CK 1542	3,252.66	
		1	04/08/85	FLA LUMBER CK 1609	142.50	
	2 ENTRIES FOR THIS TYPE			SUB-TOTALS	3,395.16	0.0
	2 ENTRIES FOR THIS COST CODE			TOTALS	3,395.16	0.0
014-108	ROUGH CARP - BENCHES					
		1	02/08/85	PAN AM LUMB CK 1477	162.78	
		1	06/10/85	FLA LUMBER CK 1693	294.69	
		1	08/08/85	FLA LUMBER CK 1761	46.09	
	3 ENTRIES FOR THIS TYPE			SUB-TOTALS	503.56	0.0
	3 ENTRIES FOR THIS COST CODE			TOTALS	503.56	0.0
014-110	ROUGH CARP - MISC NAILERS					
		1	10/15/84	MIRON CK 1048	68.32	
		1	11/09/84	MIRON CK 1147	44.45	
	2 ENTRIES FOR THIS TYPE			SUB-TOTALS	112.77	0.0
	2 ENTRIES FOR THIS COST CODE			TOTALS	112.77	0.0
014-112	SLIDING MIRROR & DOOR SUPPORT					
		1	11/09/84	NAT'L TOOL CK 1144	30.00	
		1	11/09/84	PAN AM LUMB CK 1145	57.42	
		1	11/09/84	MIRON CK 1147	61.74	
	3 ENTRIES FOR THIS TYPE			SUB-TOTALS	149.16	0.0
	3 ENTRIES FOR THIS COST CODE			TOTALS	149.16	0.0
014-114	ROOF OVER STAIR #5					
		1	02/08/85	PAN AM LUMB CK 1477	33.58	
	1 ENTRY FOR THIS TYPE			SUB-TOTALS	33.58	0.0
	1 ENTRY FOR THIS COST CODE			TOTALS	33.58	0.0
014-115	SAFEWAY EXTRA - (SCAFFOLDS)					
		1	02/08/85	PAN AM LUMB CK 1477	88.00	
		9	02/08/85	PATENT SCAFFOLD CK 14	97.34	
	1 ENTRY FOR THIS TYPE			SUB-TOTALS	88.00	0.0
	1 ENTRY FOR THIS TYPE			SUB-TOTALS	97.34	0.0
	0 ENTRIES FOR THIS TYPE			SUB-TOTALS	0.00	0.0

RUN DATE: 25-JUL-88
JOB NAME :
JOB COST CUMULATIVE TRANSACTION REPORT PAGE 91

COST-CODE	DESCRIPTION	TYPE	DATE	REFERENCE	$ AMOUNT	VOL AMOUNT
014-117	2 ENTRIES FOR THIS COST CODE VIEWING PLAZA			TOTALS	185.34	0.0
		1	06/10/85	1 FLA LUMBER CK 1693	1,573.43	
	1 ENTRY FOR THIS TYPE			SUB-TOTALS	1,573.43	0.0
	0 ENTRIES FOR THIS TYPE			TOTALS	1,573.43	0.0
	1 ENTRY FOR THIS COST CODE					
015-101	PROJECT SUPERINTENDANT	8	05/03/84	8 ZANE PLUMLEY W/E 7/31	1,250.00	
		8	05/03/84	8 ZANE PLUMLEY 8/83	6,250.00	
		8	05/03/84	8 ZANE PLUMLEY 9/83	6,250.00	
		8	05/03/84	8 ZANE PLUMLEY 10/83	6,250.00	
		8	05/03/84	8 ZANE PLUMLEY 11/83	6,250.00	
		8	05/03/84	8 ZANE PLUMLEY 12/83	6,250.00	
		8	05/03/84	8 DENNIS ALMENDARES 1/8	4,166.67	
	7 ENTRIES FOR THIS TYPE			SUB-TOTALS	36,666.67	0.0
	7 ENTRIES FOR THIS COST CODE			TOTALS	36,666.67	0.0
015-107	SUPER - CLERK OF WORKS	8	07/19/83	8 PETTY CASH	66.00	
	1 ENTRY FOR THIS TYPE			SUB-TOTALS	66.00	0.0
	0 ENTRIES FOR THIS TYPE			TOTALS	66.00	0.0
	1 ENTRY FOR THIS COST CODE					
016-104	MISC - ROUGH HARDWARE	1	08/02/83	1 PETTY CASH	4.20	
		1	08/04/83	1 BC BUILDERS	1,631.59	
		1	08/04/83	1 MIRON	9.70	
		1	09/12/83	1 BC BUILDERS	80.04	
		1	09/12/83	1 JAR SUPPLIES	57.70	
		1	09/12/83	1 WALTON-CONTEXT	81.90	
		1	10/10/83	1 BC BUILDERS	228.67	
		1	10/10/83	1 JAR SUPPLIES	3.68	
		1	10/10/83	1 MIRON	76.70	
		1	11/10/83	1 BC BUILDERS	249.93	
		1	11/10/83	1 MIRON	55.13	
		1	11/10/83	1 NATIONAL TOOL	35.28	
		1	11/10/83	1 WALTON-CONTEXT	5.99	
		1	11/10/83	1 JOHN ABELL CORP.	133.30	
		1	11/16/83	1 PETTY CASH	1.04	
		1	12/09/83	1 BAKER TOOL	45.73	
		1	12/09/83	1 BC BUILDERS HARDWARE	261.44	
		1	12/09/83	1 MIRON BUILDING PRODS.	82.51	
		1	12/09/83	1 NATIONAL TOOL SUPPLY	84.69	
		1	12/09/83	1 THOMAS MACHINERY INC.	105.69	
		1	01/10/84	1 B. C. BLDRS HDWE - CK	203.11	
		1	01/10/84	1 BLANCHARD MACHINERY -	136.6.	
		1	01/10/84	1 FLA LEVEL CK 3343	18.90	
		1	01/10/84	1 MIRON BLDG PROD CK	252.67	

RUN DATE: 25-JUL-88   JOB COST CUMULATIVE TRANSACTION REPORT   PAGE 95

JOB NAME :

COST-CODE	DESCRIPTION	TYPE	DATE	REFERENCE	$ AMOUNT	VOL AMOUNT
		1	06/10/85	M.G. ENT CK 1687	19.17	
		1	06/18/85	P/C CK 4522	50.96	
		1	06/28/85	P/C CK 4549	151.20	
		1	07/03/85	P/C CK 4561	74.72	
		1	07/10/85	NAT'L TOOL CK 1718	37.80	
		1	07/10/85	AAA TOOL CK 1715	4.50	
		1	07/17/85	P/C CK 4584	45.68	
		1	07/23/85	P/C CK 4591	52.51	
		1	07/26/85	P/C SHEET 7/26/85	32.37	
		1	08/08/85	GRABBER CK 1758	61.37	
		1	08/16/85	P/C SHEET 8/16/85	29.14	
				SUB-TOTALS	25,235.29	0.0
				TOTALS	25,235.29	0.0
016-105	MISC - SOIL POISONING AND					
		1	10/19/84	P/C CK 3962	7.28	
		1	10/19/84	P/C CK 3962	1.90	
				SUB-TOTALS	9.18	0.0
		9	09/09/83	ACTIVE PEST CONTROL	350.00	
		9	10/10/83	ACTIVE PEST	420.00	
		9	10/10/83	BC BUILDERS	11.83	
		9	10/10/83	JOHN ABELL CORP	31.48	
		9	02/10/84	ACTIVE PEST - CHECK #	542.50	
		9	03/10/84	ACTIVE PEST CONT - CK	210.00	
		9	05/10/84	ACTIVE PEST CK 538	210.00	
		9	10/15/84	ACTIVE CK 1040	349.50	
		9	07/10/85	ACTIVE PEST CONT CK 1	262.12	
				SUB-TOTALS	2,393.43	0.0
				SUB-TOTALS	0.00	0.0
				TOTALS	2,402.61	0.0
016-107	REPAIRS					
		1	04/10/84	MIRON CK 474	5.31	
		1	05/10/84	BC BLDRS CK 541	59.58	
		1	05/10/84	MIRON CK 551	52.50	
		1	06/04/84	P/C CK 622	93.45	
		1	06/08/84	MIRON CK 690	233.88	
		1	07/10/84	HILTI CK 752	327.29	
		1	10/03/84	P/C CK 1015	7.57	
		1	05/09/85	ABC CUTTING CK 1674	90.00	
		1	06/18/85	P/C CK 4522	126.00	
		1	10/10/85	TARMAC FLORIDA CK 472	88.58	
				SUB-TOTALS	1,083.16	0.0
		2	08/22/84	56772 - SPLIT		1.0
				SUB-TOTALS	0.00	1.0
		3	05/10/84	BLANCHARD CK 540	170.62	
		3	05/10/84	PHN CK 549	105.63	

188 ENTRIES FOR THIS TYPE
188 ENTRIES FOR THIS COST CODE

2 ENTRIES FOR THIS TYPE

9 ENTRIES FOR THIS TYPE

0 ENTRIES FOR THIS TYPE
11 ENTRIES FOR THIS COST CODE

10 ENTRIES FOR THIS TYPE

1 ENTRY FOR THIS TYPE

273

RUN DATE: 25-JUL-88
JOB NAME :

JOB COST CUMULATIVE TRANSACTION REPORT

COST-CODE	DESCRIPTION	TYPE	DATE	REFERENCE	$ AMOUNT	VOL AMOUNT
	29 ENTRIES FOR THIS TYPE	1	06/10/85	1 M.G. ENT CK 1687	60.70	0.0
				SUB-TOTALS	11,340.71	
		9	11/09/84	9 B.E. HENSLEY CK 1191	900.91	
	1 ENTRY FOR THIS TYPE			SUB-TOTALS	900.91	0.0
	0 ENTRIES FOR THIS TYPE			TOTALS	12,241.62	0.0
016-110	30 ENTRIES FOR THIS COST CODE BALCONY RAILINGS			**********	**********	
		1	11/09/84	1 MIRON CK 1147	31.84	
	1 ENTRY FOR THIS TYPE			SUB-TOTALS	31.84	
		3	10/15/84	3 BLANCHARD CK 1042	46.99	
	1 ENTRY FOR THIS TYPE			SUB-TOTALS	46.99	0.0
	0 ENTRIES FOR THIS TYPE			SUB-TOTALS	0.00	0.0
016-111	2 ENTRIES FOR THIS COST CODE SCUPPERS			TOTALS	78.83	0.0
		1	03/12/85	1 P/C CK 1556	74.43	
		1	03/12/85	1 P/C CK 1556	4.32	
		1	04/08/85	1 M.G. ENT CK 1603	141.75	
	3 ENTRIES FOR THIS TYPE			SUB-TOTALS	220.50	0.0
016-113	3 ENTRIES FOR THIS COST CODE TRASH CHUTE - RELATED TO			TOTALS	220.50	0.0
		1	05/09/85	1 NAT'L TOOL CCK 1656	29.65	
	1 ENTRY FOR THIS TYPE			SUB-TOTALS	29.65	
016-114	1 ENTRY FOR THIS COST CODE REPAIR TO BALCONY RAILINGS			TOTALS	29.65	
		1	05/20/85	1 P/C CK 4431	11.33	
		1	06/04/85	1 P/C CK 4468	25.20	
		1	06/04/85	1 P/C CK 4468	43.09	
		1	06/18/85	1 P/C CK 4522	11.66	
		1	07/10/85	1 SUNSHINE CONT CK 1720	141.75	
	5 ENTRIES FOR THIS TYPE			SUB-TOTALS	233.03	0.0
016-115	5 ENTRIES FOR THIS COST CODE ELEVATOR PERMITS			TOTALS	233.03	0.0
		1	05/17/85	1 ALL FLA ELEV CK 4427	600.00	
	1 ENTRY FOR THIS TYPE			SUB-TOTALS	600.00	0.0
016-116	1 ENTRY FOR THIS COST CODE PROF FEES & RELATED COSTS			TOTALS	600.00	0.0
		1	07/17/85	1 PC 4584 - PHOTOS	127.00	
		1	07/17/85	1 PC 4584 - PHOTOS	125.00	
		1	07/26/85	1 PC - PHOTOS	6.64	
		1	08/08/85	1 CIR BL 1757 - SEPIA	15.75	
		1	08/08/85	1 T-SQ 1760 - PRINTS	22.05	

PAGE 97

RUN DATE: 25-JUL-88   JOB COST CUMULATIVE TRANSACTION REPORT   PAGE 99

JOB NAME :

COST-CODE	DESCRIPTION	TYPE	DATE	REFERENCE	$ AMOUNT	VOL AMOUNT
		1	09/12/83	LONESTAR - CHECK #301	98,882.29	
		1	10/10/83	LONESTAR - CHECK #314	51,480.42	
		1	10/11/83	LONESTAR - CHECK #315	345.05	
		1	11/10/83	LONESTAR - CHECK #325	65,120.21	
		1	12/09/83	LONESTAR - CHECK #347	78,809.68	
		1	01/10/84	LONESTAR - CHECK #332	116,480.13	
		1	02/10/84	LONESTAR - CHECK #18	142,387.89	
		1	03/10/84	LONESTAR - CHECK #298	11,475.30	
		1	04/10/84	LONESTAR CK 469	82,298.92	
		1	05/10/84	LONESTAR CK 547	55,596.21	
		1	06/08/84	LONESTAR CK 687	35,797.14	
		1	07/10/84	LONESTAR CK 757	23,210.17	
		1	08/10/84	LONESTAR CK 837	36,544.69	
		1	09/10/84	LONESTAR CK 919	15,408.21	
		1	10/15/84	LONESTAR CK 1047	17,640.74	
		1	11/09/84	LONESTAR CK 1143	15,679.86	
		1	12/10/84	LONESTAR CK 1243	4,402.65	
		1	01/10/85	LONESTAR FLA CK 1372	2,277.08	
		1	02/08/85	LONESTAR CK 1464	1,156.32	
		1	02/08/85	TARMAC FLA CK 1474	3,054.97	
		1	03/10/85	TARMAC FLA CK 1540	6,850.87	
		1	04/08/85	TARMAC CK 1607	5,077.92	
		1	05/09/85	TARMAC CK 1670	1,610.74	
		1	06/10/85	TARMAC FLA CK 1691	1,279.69	
		1	07/10/85	TARMAC, INC CK 1723	229.69	
				SUB-TOTALS	1,023,428.55	0.0

27 ENTRIES FOR THIS TYPE

				SUB-TOTALS	0.00	0.0
				TOTALS	1,023,428.55	0.0

017-127   0 ENTRIES FOR THIS TYPE
           27 ENTRIES FOR THIS COST CODE
LABOR - $ VOLUME

		10	10/15/75	MED INS CK 4740 PER M	231.85	
		10	06/28/83	PAYROLL W/E 6/26/83	1,786.60	
		10	06/29/83	CARP. UNION CHECK #25	12.24	
		10	07/05/83	PAYROLL W/E 7/3/83	2,971.33	
		10	07/07/83	CARPENTERS UNION CHEC	48.96	
		10	07/12/83	CARPENTERS UNION CHEC	110.16	
		10	07/12/83	LABORERS UNION CHEC	57.33	
		10	07/12/83	LAB. PENSION - CHECK	32.83	
		10	07/12/83	PAYROLL W/E 7/10/83	2,538.53	
		10	07/14/83	CARPENTERS UNION CHEC	122.40	
		10	07/14/83	LABORERS UNION CHEC	76.05	
		10	07/14/83	LAB PENSION - CHECK #	43.55	
		10	07/19/83	Z.P. - TIMEKEEPER-CHE	66.00	
		10	07/19/83	PAYROLL W/E 7/19/83	4,789.94	
		10	07/20/83	CARPENTERS UNION CHEC	122.40	
		10	07/20/83	LABORERS UNION CHEC	94.77	
		10	07/20/83	LAB. PENSION - CHECK	54.27	
		10	07/26/83	PAYROLL W/E 7/24/83	5,181.04	
		10	08/02/83	CARPENTERS UNION CHEC	140.76	
		10	08/02/83	LABORERS UNION CHEC	81.90	

RUN DATE: 25-JUL-88

JOB COST CUMULATIVE TRANSACTION REPORT

PAGE 100

JOB NAME :

COST-CODE	DESCRIPTION	TYPE	DATE	REFERENCE	$ AMOUNT	VOL AMOUNT
		10	08/02/83	10 LAB. PENSION - CHECK	46.90	
		10	08/02/83	10 PAYROLL W/E 7/31/83	5,818.21	
		10	08/02/83	10 Z.P. - DAY LABOR -RAF	68.90	
		10	08/02/83	10 CARPNTERS UNION CHEC	122.40	
		10	08/02/83	10 LAB. PENSION - CHECK	66.66	
		10	08/02/83	10 TROWEL TRADES - CHECK	34.85	
		10	08/02/83	10 OP. ENG. LOCAL 487 CH	242.50	
		10	08/02/83	10 CARPENTERS UNION CHEC	126.99	
		10	08/02/83	10 LABORERS UNION CHECK	269.68	
		10	08/02/83	10 LAB. PENSION - CHECK	154.44	
		10	08/02/83	10 TROWEL TRADES CHECK #	52.28	
		10	08/07/83	10 LABORERS UNION CHECK	116.42	
		10	08/09/83	10 PAYROLL W/E 8/7/83	5,516.11	
		10	08/16/83	10 PAYROLL W/E 8/14/83	3,578.75	
		10	08/16/83	10 CARPENTERS UNION CHEC	122.40	
		10	08/16/83	10 LABORERS UNION CHEC	169.65	
		10	08/16/83	10 LAB PENSION - CHECK #	97.15	
		10	08/23/83	10 PAYROLL W/E 8/21/83	4,042.79	
		10	08/23/83	10 CARPENTERS UNION CHEC	122.40	
		10	08/23/83	10 LABORERS UNION CHECK	188.95	
		10	08/23/83	10 LAB PENSION - CHECK #	108.21	
		10	08/29/83	10 PAYROLL W/E 8/28/83	4,224.34	
		10	08/31/83	10 CARPENTERS UNION CHEC	122.40	
		10	08/31/83	10 LAB. PENSION - CHECK	202.41	
		10	08/31/83	10 LABORERS UNION CHEC	115.91	
		10	08/31/83	10 TROWEL TRADES CHECK #	49.20	
		10	08/31/83	10 OP. ENG. LOCAL - CHEC	33.13	
		10	09/06/83	10 PAYROLL W/E 9/4/83	9,621.15	
		10	09/12/83	10 CARPENTERS UNION CHEC	122.40	
		10	09/12/83	10 LABORERS UNION CHECK	138.06	
		10	09/12/83	10 LAB. PENSION - CHECK	79.06	
		10	09/13/83	10 PAYROLL W/E 9/11/83	5,746.90	
		10	09/20/83	10 PAYROLL W/E 9/18/83	5,015.84	
		10	09/21/83	10 CARPENTERS UNION CHEC	218.21	
		10	09/21/83	10 LABORERS UNION CHECK	124.96	
		10	09/21/83	10 LAB PENSION - CHECK #	116.85	
		10	09/21/83	10 TROWEL TRADES CHECK #	122.40	
		10	09/23/83	10 LABORERS UNION CHECK	227.57	
		10	09/23/83	10 LAB. PENSION - CHECK	130.32	
		10	09/23/83	10 TROWEL TRADES CHECK #	85.08	
		10	09/27/83	10 PAYROLL W/E 9/25/83	3,526.37	
		10	10/04/83	10 PAYROLL W/E 10/2/83	10,552.03	
		10	10/04/83	10 CARPENTERS UNION CHEC	122.40	
		10	10/04/83	10 LABORERS UNION CHECK	215.87	
		10	10/04/83	10 LAB. PENSION - CHECK	123.62	
		10	10/04/83	10 TROWEL TRADES CHECK #	125.05	
		10	10/04/83	10 CARPENTERS UNION CHEC	122.40	
		10	10/04/83	10 LABORERS UNION CHEC	122.98	
		10	10/04/83	10 LAB. PENSION - CHECK	112.23	
		10	10/04/83	10 TROWEL TRADES CHECK #	152.73	

```
RUN DATE: 25-JUL-88                    JOB COST CUMULATIVE TRANSACTION REPORT                           PAGE 110
JOB NAME :

   COST-CODE   DESCRIPTION          TYPE   DATE      REFERENCE                        $ AMOUNT      VOL  AMOUNT

                                     10   10/01/85   10 MONTHLY P/R PER MM           13,334.00
                                     10   11/14/85   10 MED INS 11/85                   290.50
                                     10   12/02/85   10 P/R MONTHLY 11/85            17,918.00
                                     10   12/04/85   10 MED INS                         290.50
                                     10   12/31/85   10 P/R MONTHLY 12/85-FIN        17,918.00
                                     10   02/03/86   10 P/R 1/86                      4,166.67
                                     10   04/01/86   10 P/R M/E 3/31 E.H. PER         4,166.67
                                                                    SUB-TOTALS    1,457,696.75         0.0
                                                                       TOTALS     1,457,696.75         0.0
   ***************************************************************************************************
   017-128     PAYROLL TAXES
                                     10   06/30/83   10 P/R TAXES 6/83                  119.71
                                     10   07/31/83   10 P/R TAXES 7/83                1,037.26
                                     10   08/31/83   10 P/R TAXES 8/83                1,439.43
                                     10   09/30/83   10 P/R TAXES 9/83                1,163.67
                                     10   10/31/83   10 P/R TAXES 10/83               1,196.45
                                     10   11/30/83   10 P/R TAXES 11/83               2,504.84
                                     10   12/31/83   10 P/R TAXES 12/83               3,389.14
                                     10   01/31/84   10 P/R TAXES 1/84                4,232.47
                                     10   02/28/84   10 P/R TAXES 2/84                6,092.11
                                     10   03/31/84   10 P/R TAXES 3/84                4,366.68
                                     10   04/30/84   10 P/R TAXES 4/84                4,528.96
                                     10   05/31/84   10 P/R TAXES 5/84                5,451.15
                                     10   06/30/84   10 P/R TAXES 6/84                3,028.90
                                     10   07/31/84   10 P/R TAXES 7/84                4,249.00
                                     10   08/31/84   10 P/R TAXES 8/84                3,609.86
                                     10   09/30/84   10 P/R TAXES 9/84                2,647.97
                                     10   10/31/84   10 P/R TAXES 10/84               5,104.99
                                     10   11/30/84   10 P/R TAXES 11/84               3,983.82
                                     10   12/31/84   10 P/R TAXES 12/84               3,334.02
                                     10   01/31/85   10 4TH QTR P/R TAXES - S         1,079.01
                                     10   01/31/85   10 P/R TAXES 1/85                3,789.78
                                     10   02/28/85   10 P/R TAXES 2/85                3,773.78
                                     10   03/31/85   10 P/R TAXES 3/85                3,148.64
                                     10   04/30/85   10 1ST QTR FUTA - 1985             884.60
                                     10   04/30/85   10 1ST QTR STATE U/C - 1         2,322.09
                                     10   04/30/85   10 P/R TAXES 4/85                2,842.14
                                     10   05/31/85   10 FICA LIAB MAY 1985            2,182.22
                                     10   06/30/85   10 FICA LIAB JUNE 1985           1,287.59
                                     10   07/31/85   10 FICA LIAB JULY 1985             911.80
                                     10   07/31/85   10 2ND QUARTER 1985 - FU           140.77
                                     10   07/31/85   10 2ND QUARTER 1985 - ST           369.51
                                     10   08/31/85   10 FICA LIAB - AUG 1985            189.42
                                     10   10/31/85   10 3RD QTR UNEMP TAX               10.29
                                     10   10/31/85   10 3RD QTR FED UNEMP TAX            3.92
                                                                    SUB-TOTALS       84,496.57
                                                                       TOTALS        84,496.57
   ***************************************************************************************************
   017-129     INSURANCE
                                     10   01/16/84   10 GCA INS - CHECK #3365         8,320.95         0.0
                                     10   04/16/84   10 GCA INS - CK 3473             5,017.00         0.0

             537 ENTRIES FOR THIS TYPE
             537 ENTRIES FOR THIS COST CODE

              34 ENTRIES FOR THIS TYPE
              34 ENTRIES FOR THIS COST CODE
```

RUN DATE: 25-JUL-88   JOB COST CUMULATIVE TRANSACTION REPORT   PAGE 111
JOB NAME :

COST-CODE	DESCRIPTION	TYPE	DATE	REFERENCE	$ AMOUNT	VOL AMOUNT
		10	06/26/85	10 GCA CK 4544 W/C INS -	3,146.00	
		10	08/07/85	10 GCA CK 4619 - GEN LIA	5,767.00	
		10	10/28/85	10 GCA INS CK 4755-W/C	35,903.00	
		10	10/28/85	10 CIGNA CK RCD REF W/C	21,190.00-	
	6 ENTRIES FOR THIS TYPE			SUB-TOTALS	36,963.95	0.0
	0 ENTRIES FOR THIS TYPE					
	5 ENTRIES FOR THIS COST CODE			TOTALS	36,963.95	0.0
017-130	PETTY CASH - PER FM			**************************		
		1	03/01/84	1 CHECK #250	1,500.00	
		1	09/18/84	1 CK 3884 PER F.M.	1,500.00	
	2 ENTRIES FOR THIS TYPE			SUB-TOTALS	3,000.00	0.0
	2 ENTRIES FOR THIS COST CODE			TOTALS	3,000.00	0.0
017-131	CPM SCHEDULE			**************************		
		1	12/10/84	1 CIRCLE CK 1238	6.30	
	1 ENTRY FOR THIS TYPE			SUB-TOTALS	6.30	0.0
		9	08/26/83	9 CMS INT'L CK 2934	7,889.67	
		9	04/13/84	9 CMS INT'L CK 3471	1,815.00	
		9	05/10/84	9 CMS INT'L CK 557	500.00	
		9	06/08/84	9 CMS INT'L CK 694	500.00	
		9	07/10/84	9 CMS CK 763	500.00	
		9	08/10/84	9 CMS INT'L CK 848	500.00	
		9	09/10/84	9 CMS INT'L CK 925	500.00	
		9	10/15/84	9 CMS INT'L CK 1053	500.00	
		9	11/09/84	9 CMS CK 1150	500.00	
		9	12/10/84	9 CMS INTL CK 1251	500.00	
		9	01/10/85	9 CMS CK 1378	500.00	
	11 ENTRIES FOR THIS TYPE			SUB-TOTALS	13,404.67	0.0
	0 ENTRIES FOR THIS TYPE					
	12 ENTRIES FOR THIS COST CODE			TOTALS	13,410.97	0.0
017-132	PETTY CASH FUND TO JOBSITE			**************************		
		1	12/31/85	1 TO ALLOCATE P/C	300.00-	
	1 ENTRY FOR THIS TYPE			SUB-TOTALS	300.00-	0.0
		10	01/25/84	10 TO OPEN P/C FUND FOR	431.54	
		10	10/04/84	10 CK 3916 TO INC P/C FU	500.00	
		10	07/26/85	10 P/C SLIP 7/26/85	103.64-	
		10	08/16/85	10 NO CHECK MADE TO REIM	743.62-	
		10	09/06/85	10 P/C SLIP 9/6/85	397.21-	
		10	09/06/85	10 CK 4661 - D.A. TO REI	312.93	
		10	09/06/85	10 CK 4662 TO DA FOR P/C	300.00	
	7 ENTRIES FOR THIS TYPE			SUB-TOTALS	300.00	0.0
	8 ENTRIES FOR THIS COST CODE			TOTALS	0.00	0.0

RUN DATE: 25-JUL-88   JOB COST CUMULATIVE TRANSACTION REPORT   PAGE 112

JOB NAME :

COST-CODE DESCRIPTION   TYPE   DATE   REFERENCE $ AMOUNT   VOL AMOUNT

4,637 TOTAL ENTRIES   BUDGET $ 0.00   PERCENTAGE $ 104.14   3,803,604.22
                           BUDGET VOL   28,821   PERCENTAGE VOL   30,015.0

Index

Accountant relations, 26
Accounting, 21
   and bookkeeping, 22
   and equipment purchases, 22–23
   preparing cash flow projections, 23
   and tax returns, 21–22
Accrual method of reporting of taxable income, 22
Agent of owner:
   relationship with project manager, 69–70
   responsibilities of, 43–46
     in change order work, 47–48
     in color and material selections, 47
     in construction documents, 46–47
     in inspections and testing, 48
     in meetings, 48
     in owner's use prior to completion, 48–49
     in requisitions for payment, 49
     in shop drawings and samples, 47
   (*See also* Owner)
AIA documents, 12
Architect, construction meetings with, 103–104
Architect job reports, 86
Architectural superintendents, 72–73
As-built drawings, 87–88, 113
Attitude of subcontractor's staff, 38
Attorney, need for, 19

Bid assembly and preparing job estimate, 53–54
Bid bond, 28
Bid price, 36
Bonding capacity of subcontractor, 37
Bonding of project, 4, 28–30
Bookkeeping, 22–23
   and accountant relations, 26

Bookkeeping (*Cont.*):
   in field office, 74
   job cost budgets, 24
   job cost management files, 23–24
   and payment requisitions, 25
   subcontractors-suppliers, 24–25
   and unions, 25–26
Brochures, 32
Builder, 1
   philosophy of, 2
   (*See also* General contractor)
Business knowledge in negotiating contracts, 11

Cash basis method of reporting of taxable income, 22
Cash flow projections, preparing, 23
Certificates of occupancy, 114
Change order work, 65–66
   coverage of potential, in construction meeting, 102
   responsibility of owner in, 47–48
Change order work files, 84
Change orders, 70, 94–95
   coverage of, in preconstruction meetings, 76–77
   follow-up, 96
   and need to update schedule, 66
   and pricing, 95–96
   problems with, 97
   processing and updating, 96–97
Clerk of the works, 74
Color selections, responsibility of owner in, 47
Commercial projects, 7
Completed contract method of reporting of taxable income, 22
Completion, performance after, 116

Computer:
  in comparing cost control records, 83
  in compiling work lists, 111
  in estimating, 53
  in scheduling, 65
Condominiums and profit/loss potential, 5
Conflict of work items and value engineering changes, 59–60
Construction commencement, 99–100
  coverage of, in preconstruction meetings, 77
Construction company, philosophy of, 2
Construction documents, responsibility of owner in, 46–47
Construction industry:
  new locales for, 118–120
  potential for growth in, 117–124
Construction management, 120–123
  equity positions, 123–124
Construction meetings, 100
  with contractor staff, architect, and engineers, 103–104
  with contractor staff and owners, 104
  with contractor's and subcontractor's staff, 103
  with general contractor staff, 100–101
    general job strengths and weaknesses, 102–103
    job cost, 101
    job progress schedule, 101
    owner-caused delays and problems, 101–102
    potential change order work, 102
    subcontractor performance, 101
Contract law, need for knowledge of, 12
Contractor (see General contractor)
Contracts:
  cost plus fee based on percentage of job cost, 10
  cost plus fixed fee, 9–10
  execution of, 12–13
  financing, 18
  lump sum or fixed price, 9
  negotiating, 10–12
  preparation and review of, 18
  upset price, 10
Corrective work, 41
  (See also Punch list work)
Cost budgets, 76
  coverage of, in preconstruction meetings, 76–77
  and preparing job estimate, 54–55

Cost budgets (Cont.):
  review of, by project manager, 70
Cost control records, 83
Cost estimates and quality control, 107
Cost plus fee based on percentage of job cost contract, 10
Cost plus fixed fee contracts, 9–10

Design consultants and value engineering changes, 60–61

Electrical superintendents, 73
Engineer job reports, 86
Engineers, construction meetings with, 103–104
Equipment and need for accountant's advice, 22–23
Equity positions, 123–124
Estimating, 51–52
  bid assembly, 53–54
  cost budgets, 54–55
  final review and adjustments, 54
  problems in, 55
  proposal forms, 54
  quantity estimates, 52–53
  recap and spreadsheets, 52
  and review of plans and documents, 52

Field offices, 34
Field production files, 84
  architect and engineer job reports, 86
  as-built drawings, 87–88
  change order work files, 84
  estimating files, 84
  general correspondence:
    contractor-owner-consultants, 85
    general contractor-subcontractors, 85
  guarantees, 87–88
  inspection reports, 85–86
  job logs, 86
  job progress schedules, 85
  minutes of meetings, 87
  progress reports, 87
  project contract, 84
  punch list data, 87
  requisitions, 88–89
  shop drawings and submittal data, 86–87
  subcontractor agreements, 84–85
  test reports, 86
  warranties, 87–88
Field supervision, 67–69
  architectural superintendents, 72–73

Field supervision (*Cont.*):
coverage of, in preconstruction meetings, 76
electrical superintendents, 73
field office support staff, 74
bookkeeper, 74
clerk of the works, 74
secretary, 74
interior design superintendents, 73
labor superintendents, 73–74
mechanical superintendents, 73
plumbing superintendents, 73
project manager, 69
budgets and records, 70
change orders, 70
owners and owner agents, 69–70
progress schedules, 70
requisitions for payment, 70–71
shop drawings, support data, and samples, 70
project superintendent, 71
assistants' responsibilities, 71
coordination with project manager, 71
and job morale, 72
subcontractor work, 71–72
and work performed directly by general contractor, 72
punch list superintendents, 73
structural superintendents, 72
Final documents:
as-built drawings, 113
certificates of occupancy, 114
warranties and guarantees, 113–114
Final review and adjustments and preparing job estimate, 54
Final waiver of lien, 131–132
Finances and need for attorney, 18
Financial viability:
of project, 4
of subcontractor, 37
Fixed price contract, 9
and renovations, 7
Float time, 66
Forms, coverage of, in preconstruction meetings, 80
Formtool, 111

General contractor, 1
bonding capacity of, 4
and competition, 5
and coverage of work performed in preconstruction meetings, 76
General contractor (*Cont.*):
and emotional commitment, 5
emotional requirements of, 1–2
need for legal knowledge, 12
need for understanding owner, 12
opening offices in new locales, 118–120
and quality control, 106–107
review of shop drawings and submittal data by, 94
staff availability, 4–5
versatility of, 3–4
work performed by, and role of project superintendent, 72
General contractor staff, construction meetings with, 100–104
General counseling and need for attorney, 19–20
Government-municipal projects, 6
and profit/loss potential, 5
Guarantees, 87–88, 113–114

High rises, 6

Industrial buildings, 6
Inspection reports, 85–86
Inspections and testing, responsibility of owner in, 48
Insurance, 27–28
bonds as form of, 28–30
and need for attorney, 19
Interior design superintendents, 73

Job cost, coverage of, in construction meeting, 101
Job cost budgets, 24, 54–55
Job cost management files, 23–24
Job cost management printout:
budget and actual costs, 165–210
volume and dollar amounts, 211–279
Job logs/records, 86
coverage of, in preconstruction meetings, 76
preparation of, 70
Job morale, role of project superintendent in maintaining, 72
Job progress:
and need to update schedule, 66
schedules of, 38–39
Job progress schedules, 38–39, 63–65, 76, 85
belief in, 65
change order work, 65–66
correcting errors in, 65

Job progress schedules (*Cont.*):
coverage of:
in construction meeting, 101
in preconstruction meetings, 76, 78–79
graphics in, 64
lead time in, 63
and meetings with other principles, 64
preparation of, 63–64
and project manager, 70
structure and composition of, 63
updating, 66
worth of, 63
Job strengths and weaknesses, coverage of, in construction meeting, 102–103

Labor law and need for attorney, 19
Labor superintendents, 73–74
Lead time in job progress schedule, 63
Legal matters, 15–18
company structure, 18
financing, 18
general counseling, 19–20
insurance, 19
labor law, 19
litigation, 19
and need to understand contract law, 12
negotiations, 19
preparation and review of contracts, 18
subcontractor agreements, 18–19
Letters of reference, 33
Lien:
definition of, 24
partial release of, 129
release of, 24–25
waiver of, 130
Litigation and need for attorney, 19
Logs, 93
Lump sum contracts, 9

Main offices, 34
Material selections, responsibility of owner in, 47
Mechanical superintendents, 73
Meetings, responsibility of owner in, 48
(*See also* Construction meetings; Preconstruction meetings)

Negotiation:
of contracts, 10–12
and need for attorney, 19

Owner:
and construction documents, 46–47
construction meetings with, 104
coverage of delays and problems caused by, in construction meeting, 101–102
coverage of special requests of, in preconstruction meetings, 77–78
relationship with project manager, 69–70
responsibilities of, 43–46
in change order work, 47–48
for color and material selection, 47
in inspections and testing, 48
in meetings, 48
in requisitions for payment, 49
in shop drawings and samples, 47
in use prior to completion, 48–49
unreasonable requests of, and quality control, 107–108

Partial release of lien, 129
Payment requisitions, 25
coverage of, in preconstruction meetings, 80
and need to update schedule, 66
Payment(s):
final checking, 115
owner payments to subcontractors and suppliers, 115
support data, 115
Percentage of completion method of reporting of taxable income, 22
Performance and payment bonds, 28–30
Photographs, 33
Plumbing superintendents, 73
Preconstruction meetings:
first, 75
commencement of construction, 77
cost budgets, 76–77
field supervision responsibilities, 76
forms, records, and logs, 76
job progress schedules, 76
plans, specifications, and addenda, 75
subcontractors and suppliers, 76
work performed by general contractor, 76
second:
change orders, shop drawings, and requisitions, 77
introduction, 77
job progress schedule, 78

Preconstruction meetings, second (*Cont.*):
  owner-furnished items, 77–78
  owner's special requests, 78
  and value engineering changes, 78
  third, 78–79
    clarifications, 79
    forms, 80
    job progress schedule, 79
    payment requisitions, 80
    shop drawings and submittal data, 79–80
    storage and parking areas, 80
    subcontract agreements, 79
  fourth, 80–81
Pricing of change orders, 95–96
Progress reports, 87
Project contract, 84
  (*See also* Contracts)
Project manager:
  budgets and records, 70
  change orders, 70
  coordination with project superintendent, 71
  owners and their agents, 69–70
  progress schedules, 70
  requisitions for payment, 70–71
  shop drawings, support data, and samples, 70
Project signs, 31–32
Project superintendent, 71
  assistants' responsibilities, 71
  coordination with project manager, 71
  and job morale, 72
  subcontractor work, 71–72
  and work performed directly by general contractor, 72
Projects:
  bonding of, 4, 28–30
  competition, 5
  emotional commitment, 5
  financial viability of, 4
  profit/loss potential, 5
  selection of, 3–4
  staff availability for, 4–5
  types of, 5–7, 117–118
    commercial, 7
    government-municipal projects, 6
    high rises, 6
    industrial buildings, 6
    renovations, 7
    residential work, 7
Proposal forms and preparing job estimate, 54

Public relations, 31
  brochures, 32
  field offices, 34
  letters of reference, 33
  main offices, 34
  photographs, 33
  project signs, 31–32
  publications, 33
Publications, 33
Punch list data, 87
Punch list superintendent, 73, 108
Punch list work, 41
  follow-up, 111–112
  and quality control, 105–108
  and subcontractors, 110–111
  supervision of, 108
  when to start, 109–110
  work lists, 111

Quality control, 105–106
  and contractor-performed work, 106–107
  and cost estimates, 107
  and punch list work, 108–112
  and subcontractors, 39–40, 106
  and unreasonable owner requests, 107–108
Quantity estimates and preparing job estimate, 52–53

Recap and spreadsheets and preparing job estimate, 52
Records and notices and value engineering changes, 60
Release of lien, 24–25
Renovations, 7
Request for payment, 125
Requisition for payment, 88–89
  coverage of, in preconstruction meetings, 77
  responsibility of owners in, 49
  role of project manager in, 70–71
  for subcontractors, 40–41
Requistion printout, 153–160
Residential work, 7
Review of plans and documents and preparing job estimate, 52
Reward/risk ratio, 5

Schedule (*see* Job progress schedules)
Scopes of work, 36
Secretary of field office, 74
Shop drawings, 40, 70, 86–87, 91–93

286　Index

Shop drawings (*Cont.*):
　coverage of, in preconstruction meetings, 77, 79–80
　logs, 93
　responsibility of owner in, 47
　(*See also* Submittal data, forms for)
Specialization, advantage in, 3–4
Specialty contractors, 3
Staff availability for project, 4–5
Structural superintendents, 72
Subcontract agreement, 84–85, 133–152
　coverage of, in preconstruction meetings, 79
　and need for attorney, 18–19
Subcontractor requisition forms, 127
Subcontractor status report printout, 161–163
Subcontractors, 24–25, 35–36
　bonding capacity of, 37
　and certificates of insurance, 28
　coverage of:
　　in construction meeting, 101
　　in preconstruction meetings, 76
　financial viability, 37
　on-the-job problems and solutions:
　　job progress schedules, 38–39
　　punch list work, 41
　　requisitions for payment, 40–41
　　shop drawings and submittal data, 40
　　substandard quality, 39–40
　owner payments to, 115–116
　and payment requisitions, 25
　precontract problems and solutions:
　　attitude of staff, 38
　　bid price, 36
　　scopes of work, 36
　　staff, 36–37
　and project superintendent, 71–72
　and punch list work, 110–111
　and quality control, 106
　and value engineering changes, 37–38

Subcontractors (*Cont.*):
　workload of, 37
Submittal data, 86–87
　coverage of, in preconstruction meetings, 79–80
　forms for, 93–94
　general contract review of, 94
　timing processing and distribution, 94
Substandard quality, 39–40
Suppliers, 24–25
　owner payments to, 115–116
Support data, 70
Support staff:
　of field office, 74
　field office bookkeeper, 74
　field office clerk of the works, 74
　field office secretary, 74

Tax returns, 21–22
Taxable income, methods of reporting, 22
Technical knowledge in negotiating contracts, 11
Test reports, 86

Unions, 25–26
Upset price contracts, 10
Use prior to completion, responsibility of owner in, 48–49

Value engineering changes, 57–59
　conflict of work items, 59–60
　and design consultants, 60–61
　in negotiating contracts, 11–12
　records and notices, 60
　at second preconstruction meeting, 78
　and subcontractors, 37–38

Waiver of lien, final, 131–132
Warranties, 87–88, 113
Work lists and punch list work, 111
Work samples, 70
Workload of subcontractors, 37

About the Author

Mert Millman is founder and principal of Millman Construction Company, Inc., based in Miami, Florida. In business for over thirty years, Millman Construction has built numerous public and private projects, including convention centers, community colleges, a marine stadium, schools, shopping centers, industrial buildings, and other projects of various types and uses. In addition, Mr. Millman's Company has been responsible for constructing some 6,000 apartment and condominium units throughout South Florida.